TIMOTHY J. RYAN, Ph.D.
SCHOOL of HEALTH SCIENCES
OHIO UNIVERSITY
409 THE TOWER
ATHENS, OH 45701-2979

TIMOTHY J. RYAN, Ph. D., CIH, CSP
OHIO UNIVERSITY
E327 GROVER CENTER
ATHENS, OH 45701-2979

RESEARCH DESIGN
and STATISTICS *for the*
SAFETY *and* HEALTH
PROFESSIONAL

RESEARCH DESIGN
and STATISTICS *for the*
SAFETY *and* HEALTH
PROFESSIONAL

Charles A. Cacha, Ph.D., CSP, CPE, ARM

VAN NOSTRAND REINHOLD

I(T)P®   A Division of International Thomson Publishing Inc.

New York • Albany • Bonn • Boston • Detroit • London • Madrid • Melbourne

Mexico City • Paris • San Francisco • Singapore • Tokyo • Toronto

Copyediting: Patricia Lewis
Proofreading: Amy Mayfield
Composition and Artwork: Carlisle Communications, Ltd.
Index: Schroeder Indexing Services
Production Editor: Emily Autumn

I(T)P® Published by Van Nostrand Reinhold, an International Thomson Publishing company.
The ITP logo is a registered trademark used herein under license.

Printed in the United States of America

http://www.vnr.com   Visit us on the web!

For more information, contact:

Van Nostrand Reinhold   Chapman & Hall GmbH
115 Fifth Avenue   Pappalallee 3
New York, New York 10003   69469 Weinham
                                        Germany

Chapman & Hall   International Thomson Publishing Asia
2-6 Boundary Row   60 Albert Street #15-01
London SEI 8HN   Albert Complex
United Kingdom   Singapore 189969

Thomas Nelson Australia   International Thomson Publishing Japan
102 Dodds Street   Hirakawa-cho Kyowa Building, 3F
South Melbourne, 3205   2-2-1 Hirakawa-cho, Chiyoda-ku
Victoria, Austria   Tokyo 102 Japan

Nelson Canada   International Thomson Editores
1120 Birchmount Road   Seneca, 53
Scarborough, Ontario   Colonia Polanco
Canada M1K 5G4   11560 Mexico D.F. Mexico

1 2 3 4 5 6 7 8 9 10 **EDWAA** 02 01 00 99 98 97

Library of Congress Cataloging-in-Publication Data

Cacha, Charles A.
   Research design and statistics for the safety and health
professional / Charles A. Cacha.
      p.   cm.
   Includes index.
   ISBN 0-442-02041-4
   1. Industrial safety--Statistical methods. 2. Industrial hygiene-
-Statistical methods. 3 Research, Industrial. I. Title.
T55.3.S72C33 1997
363.11- ·dc21   96-37475
                                                          CIP

To my faithful, patient, beloved wife Frances B. Cacha, Ph.D.

CONTENTS

PREFACE

To the Reader: This book is intended to fill a void within the repertory of skills possessed by the typical safety and health professional. I have been associating with safety and health professionals and observing their behavior for more than 20 years. My impression is that great advances in our technical competence have occurred. There is, however, one area of expertise where knowledge has developed slowly. Many safety professionals feel reluctant and downright insecure about engaging in fact finding, research, analysis, and decision making based upon the collection of data. Despite this insecurity, it is imperative and unavoidable that safety research be conducted in all organizations that are seriously dedicated to diminishing trauma and disease in the workplace and public places as well.

This book is not intended to create statisticians or research specialists. It is the intent of this book to offer comprehensible step-by-step procedures providing valuable information that will improve the safety efforts of an organization. The examples and problems presented will reflect possible real-life situations in which safety problems may be defined and resolved. It should be noted that these examples, although of a realistic nature, are fictitious and have not been based upon any events or conditions found in any particular organization.

The math skills needed to understand the text and to solve the problems at the end of the chapters are at a reasonably basic level of competence. Appendix A of this book contains elementary mathematical rules and functions that should help anyone who needs a review of basic math principles. At times, problem solving may be somewhat lengthy and tedious, but the procedures provided in the book are not beyond the abilities of a determined professional.

The technical aspects of this book represent a collection of numerous personal experiences and readings in various standard textbooks. The book that may best supplement what is presented here is Murray R. Spiegel's *Statistics,* second edition, (McGraw-Hill).

In closing, I wish to extend thanks to two friends and colleagues for their assistance in creating this book. I greatly appreciate the contribution of Mr. Roger Ramsay for writing the final chapter about the use of computers in statistics and research. I also wish to thank Ms. Janet L. Signo for her word processing and organizational efforts.

I sincerely hope that this book contributes to the professional growth of all of its readers and to the growth of our profession as well.

Charles A. Cacha, Ph.D. CSP, CPE, ARM

RESEARCH DESIGN *and* STATISTICS *for the* SAFETY *and* HEALTH PROFESSIONAL

Chapter 1

Research

1.1 Knowledge. Knowledge is the ultimate resource needed for the development and support of civilizations and their technologies. Human progress has been closely associated with the possession of knowledge. Historical studies of other eras, such as the Middle Ages, have disclosed that inadequate and insufficient knowledge has led to an eclipse of human achievement. Our storehouse of knowledge has grown substantially since the Middle Ages, however. This growth can be attributed to various intellectual authorities, most of whom have proved beneficial and some of whom have been potentially detrimental.

1.1.1 The Authority of Tradition. In the past, tradition largely governed the contents of human knowledge. People knew something because it had always been known by themselves and their forebears. The authority of tradition is an acceptable agency for the growth of knowledge provided its information is valid and complete. "The sun always rises in the east and goes down in the west" is an acceptable traditional belief, but "the earth is flat" has been demonstrated to be an invalid traditional belief.

1.1.2 The Authority of Institutions and Experts. Humans have also accepted as knowledge the statements of churches, organizations, and respected individuals. As in the case of tradition, these statements have sometimes been invalid. The long-established yet invalid idea that "the sun revolves around the earth," for example, led to prolonged critical misconceptions about our universe.

1.1.3 The Authority of Reason. Knowledge may be derived from our own internal thought processes. The rational approach, which uses the process of reasoning, arrives at conclusions through premises, which are in turn based upon an accumulation of prior knowledge. Unfortunately, this rational approach has its limitations. Two individuals may reason through a problem and come to different conclusions because of false premises or faulty logic.

1.1.4 The Authority of Observation. Knowledge may also be derived from experiences based upon observations, tests, and analyses of anything and everything that is external to the human being. This approach, which is also known as the empirical approach, collects facts about the environment that are directly or indirectly observable by human sensations.

1.2 Scientific Research. Scientific research, which shall hereafter be referred to as "research," is an amalgam of the authorities of reason and observation. The object of research is to make observations that provide valid facts that will become valid premises for valid conclusions. At the outset, it must be realized that research may be very practical in nature or very impractical, and that the degree of practicality of a research project depends on individual judgment. Whether practical or impractical, good research has some important characteristics.

1.2.1 Research Should Be Objective. A research project should not be designed to favor the personal biases of the researcher. Research should be designed so that the research outcome may just as readily prove the researcher's belief (hypothesis) to be wrong as correct.

1.2.2 Research Should Be Thorough and Methodical. Good research procedure is thorough and requires constant, uniform efforts in data gathering. Research procedures should be consistently the same from the beginning to the end of a project, and no pertinent facts should be ignored.

1.2.3 Research Should Be Directed toward the Research Problem. For the sake of clarity of results, simplicity of conclusions, and economy of effort, the research design should not include extraneous procedures that are not directed toward the research problem at hand.

1.2.4 If Possible, Research Should Be Quantifiable. Research descriptions and comparisons that are reduced to numerical values are more objective in nature and allow for greater simplicity, making the research results more readily understandable.

1.2.5 Research Results Should Be Truthfully Reported. The outcome of research should be fairly and objectively reported whether or not the outcome coincides with the researcher's expectations.

1.3 Types of Research. Research may be subdivided into procedural categories. The choice of category may be a question of personal interest or current needs. Some major categories of research follow.

1.3.1 Philosophical Research. Philosophical research is concerned with the analysis and understanding of the thoughts, theories, and opinions of individuals or groups of people. By understanding the thoughts of others, researchers may be able to propose new theories of their own.

1.3.2 Historical Research. Historical research studies trends and changes over time. Like philosophical research, historical research provides an understanding of the thoughts and circumstances of individuals or groups that may lead to the formulation of new theories in the present.

1.3.3 Descriptive Research. Descriptive research observes and describes a large or small segment or segments of the environment at a single point in time, usually the present. The observation procedure may be quite simple or quite complex as in the case of tests and other analytical procedures.

1.3.4 Experimental Research. Experimental research entails (1) observation of a segment of the environment, (2) manipulation of that segment, and (3) finally, reobservation of the segment to determine if the manipulation has had any effect. The manipulation is often referred to as a "treatment." Experimental research is most often found in the physical, chemical, and biological sciences, but may also be used in the social, behavioral, and management sciences. Experimental research in these latter sciences is often referred to as "quasi-experimental" research because of the inherent difficulties and complexities of choosing, manipulating, testing, and observing human beings.

1.3.5 Examples of Descriptive versus Experimental. Two examples will help to illustrate the difference between descriptive and experimental research.

1.3.5.1 Descriptive. Researchers administer an IQ test and a questionnaire about dietary habits to a group of underprivileged children. The researchers report the results of both instruments and offer some conclusions about the children.

1.3.5.2 Experimental. (1) Researchers administer an IQ test and a questionnaire about diet to a group of underprivileged children. (2) The researchers then place the children on a highly nutritious diet for a period of time. (3) The researchers then administer the same IQ test again, compare the first and second results, and offer conclusions based upon the comparison.

1.4 Research and Disciplines. All disciplines and areas of interest are likely to use and should use research procedures. Whether research is in the basic physical and biological sciences or in the social, behavioral, and management sciences, the procedures have a strong thread of logical commonality. Nevertheless, each of these disciplines will use a unique approach and unique methods of testing and observing. Research in chemistry, for example, is not exactly the same as research in psychology even though the same correct principles of research design apply to both.

1.4.1 Safety and Health. The safety and health discipline has its own peculiar characteristics and its own unique research requirements, yet these requirements will also conform to good research design practiced by all other disciplines. Three types of research are likely to be used in safety and health: historical research, descriptive research, and experimental research. Ensuing chapters of this book will describe various techniques that may be used in safety and health research.

1.5 Research Procedure. A typical research procedure should include the following steps: (1) state the research problem; (2) construct a hypothesis that, if supported, will solve the problem; (3) design a research procedure that will provide information that either supports or does not support the hypothesis; (4) execute the research procedure in an objective, uniform, methodical manner; and (5) come to a conclusion based upon the information. An example based upon the situation described in sections 1.3.5.1 and 1.3.5.2 follows.

1.5.1 Problem. A psychologist and a nutritionist, after discussions and reading various literature, form a belief that underprivileged children display relatively low IQ scores because of poor eating habits. They believe that there is a relationship between IQ and nutrition and decide to do some research to prove whether this is true or not true.

1.5.2 Hypothesis. The researchers formally state a research hypothesis and say: "If poorly nourished underprivileged children with low IQs are provided with a well-balanced diet, then their IQs will improve after the diet has been administered." The key words *if* and *then* are frequently used by researchers to construct a logical, understandable hypothesis.

1.5.3 Procedure and Information Gathering. The researchers carefully design and execute a research procedure that will provide information that will support or not support their hypothesis. This particular design, which must be carefully, strictly, and objectively adhered to, will be an experimental design in that it will manipulate the subjects by changing their diets. Various other designs will be described in ensuing chapters.

1.5.4 Execution. The researchers execute their research design, which involves: (1) administering an IQ test to a group of underprivileged children identified as poorly nourished by a questionnaire, (2) placing them on a beneficial diet for six months, and (3) readministering the same IQ test.

1.5.5 Conclusion. The researchers state a conclusion. The conclusion is based upon the objectively reported results of the research procedure and whether the hypothesis was supported or not supported.

1.6 Research and Statistics. Quantification is an indispensable supporting tool of research procedures. This quantification largely takes the form of statistics, a branch of mathematics that deals with gathering numerical information and then often applying the laws of probability to the information. A simpler form of statistics used in elemental research designs is known as descriptive statistics. A more complex form of statistics used in more elaborate research designs is known as inferential statistics. Ensuing chapters will provide statistical tools of both varieties. This book will not attempt to convert the safety and health professional into a statistician and thus will not provide the mathematical derivations of the various statistical tools used. This book will only provide simple step-by-step statistical procedures needed to solve the research problem at hand.

Chapter 2

The Before Only Design

2.1 The Before Only Design. The before only design is the simplest, most basic, and most frequently used format in safety and health research. This form of research observes a particular segment of the environment (hereafter referred to as a *population*) and describes it, usually in numerical terms. The population of interest is most often studied in the present but may also be studied in the past, provided adequate historical records are available. An important characteristic of the before only design is that the population is not compared with other populations, although comparisons may be made within the population itself. The before only design is part of descriptive research, and the numbers used to support its conclusions are referred to as descriptive statistics. This design uses several techniques and procedures, which will be explained in this chapter. These procedures are also used in other, more complex research designs.

2.2 Sampling and Enumeration. A population is composed of a number of individual units, usually referred to as *members*. In safety and health research, members of a population are often people but may also be objects such as machines, equipment, buildings, vehicles, and the like. The population to be studied may be examined entirely or partially. An examination of an entire population is referred to as an *enumeration*. Within reason, an enumeration observes every single member of the population. An examination of only part of the population is called a *sampling*. A sample is the result of selecting only some of the population members. Generally, only smaller populations are enumerated. For practical reasons, larger populations must be sampled. In safety and health research, enumerations are often performed because smaller groups and organizations are usually involved. If samples must be generated on larger groups, however, then certain unbiased, methodical procedures must be adhered to.

2.2.1 Sampling Techniques. There are three basic sampling techniques. (1) In simple random sampling, unique consecutive numbers, usually beginning at 1, are assigned to all members of the population. A table of random numbers then determines which of the members will become part of the sample. Hence, each member of the population has an equal chance of being chosen for the sample. Appendix D of this book contains a table of random numbers and instructions on their use. (2) Stratified random sampling may be a more desirable technique if the population is subdivided into several critical characteristics whose difference may influence the outcome of the study. If a population is composed of 75% males and 25% females and the researcher believes that sexual difference is critical to the study, then the researcher should draw a simple random sample from each sex and then lump the samples together into one sample. To obtain proportionate representation in this example, the male sample must be three times as large as the female sample. (3) Cluster sampling is similar to stratified sampling except that consideration is given to administrative, political, or geographical units. If a population such as a business organization contains Department A and Department B and the researcher believes that Department A is critically different from Department B, separate proportionate random samples of each department should be drawn and then added together.

2.2.2 Sample Size. As previously indicated, enumeration is the most desirable procedure and is often feasible in safety and health research. If sampling is required, however, then the general rule of thumb is "the larger the sample size the better." Samples should contain at least 30 members. Smaller samples, and for that matter small enumerations, may be used, but because of their small size, it may be more difficult to arrive at significant conclusions.

2.3 Categories. A normal response when examining a population of persons or objects is to classify its members according to their various characteristics. Characteristics may be based upon physical, psychological, socioeconomic, or other types of criteria. Suggested categories have been devised by the American National Standards Institute (ANSI), the National Safety Council, and the Occupational Safety and Health Administration (OSHA). Chapter 14 of this book will describe these procedures further. Categorizing leads to a better understanding of the population and its members. Categories should be carefully defined so that there is no question in which category a population member belongs. There are two basic types of categories.

2.3.1 Qualitative Categories. Qualitative classifications are easily understood and are based upon well-defined "either-or" and "black and white" categories. The following are some examples:

Sex: male or female
Worker: manufacturing or clerical
Machine: guarded or unguarded

Vehicle: inspected or uninspected
Recorded injury: lost workdays or no lost workdays

These categories are often dichotomous but may be multiple such as:

Opinion: agree or disagree or no opinion
Department: Department A or Department B or Department C and so on.

Numbers known as *frequencies* are usually assigned to these categories as in the following:

Sex: 30 males and 15 females
Machines: 10 guarded and 2 unguarded
Workers: 100 manufacturing and 10 clerical

2.3.2 Quantitative Categories. Quantitative categories are defined by their position along a continuous standard measurement scale such as height in inches or age in years. A continuous scale unavoidably produces "grays" in between "blacks and whites." These measures may be converted into qualitative categories. In the question of short worker or tall worker, we might categorize height as follows: worker under 60 inches in height or worker 60 inches or more in height. In a question of young worker or middle-aged worker or old worker, we might categorize age as follows: workers less than 35 years old or workers 35 to 55 years old or workers over 55 years old. As in the prior method, numbers are assigned to these categories and are referred to as frequencies.

2.4 Percentages. The use of categories and frequencies aids greatly in understanding a population. This method may be enhanced by the use of percentages. If the category of sex is applied to a population and the researcher discovers that the population contains 180 male members and 60 female members, it is more comprehensible to say that the population is 75% male and 25% female. A percentage is derived by dividing the frequency of the characteristic in question by the total members in the category. Thus, 180 male members divided by a total of 240 members equals .75 or 75%.

2.5 Central Tendencies. In section 2.3.2, we discussed categories that are measurable on a continuous scale. These categories are sometimes called *variables* because their values readily vary in small increments from one member to another. When presented with a group of numbers on a continuous scale, the numbers will be best understood by creating one representative number, known as a *measure of central tendency*. Two major central tendencies are available to safety and health research.

2.5.1 The Mean and the Range. The well-known *mean* or "average" is the most frequently used representation of a group of numbers along a continuous

Table 2.1. Approximate Mileage of the Truck Fleet

Truck #	Miles
1	10,000
2	8,000
3	2,000
4	13,000
5	20,000
6	40,000
7	1,000
8	50,000
9	5,000
10	15,000
11	18,000
Total	182,00

scale. Let us take an example involving the vehicles in Table 2.1. A company has a fleet of 11 trucks. The safety manager is interested in the condition of these trucks. The table lists the approximate mileage of each truck.

For convenience the fleet manager rounded off the actual mileage to the closest 1,000; thus, Truck 1's mileage of 9,903 became 10,000, and Truck 2's mileage of 8,491 became 8,000 and so on. Rounding is an acceptable procedure provided it is done logically and consistently. Rounding is not mandatory, and judgment will indicate whether it is appropriate or not. To arrive at the mean, all mileages were added up and then divided by the total number of vehicles. Thus, the mean equals 182,000 divided by 11 or 16,545.5 miles, rounded to 17,000 miles. Note also that the lowest mileage is 1,000 and the highest is 50,000. This distance from the lowest to highest value is called the *range*. If asked to give an opinion about the safety and reliability of the fleet, the safety manager might say: "There are 11 trucks in the fleet. The mileage of the trucks ranges from 1,000 miles to 50,000 miles, and the average mileage is 17,000 miles." This information, plus other information available, may help the company make a decision about vehicle replacement and fleet safety. A final note: the fleet manager provided the safety manager with rounded numbers, which were averaged and then rounded again. If the fleet manager had provided exact numbers, the preferred procedure would have been to average those numbers and then round the result.

Table 2.2. Ranking the Trucks

Rank	Truck #	Mileage
1	7	1,000
2	3	2,000
3	9	5,000
4	2	8,000
5	1	10,000
6	4	13,000
7	10	15,000
8	11	18,000
9	5	20,000
10	6	40,000
11	8	50,000

2.5.2 The Median. A less frequently used measure of central tendency is the *median*, which relies upon ranking the numbers in question. Ranking is a procedure in which numbers are listed in positions relative to their magnitude. Ranking is usually done from the lowest to highest number but may also be done from highest to lowest. Table 2.2 ranks the 11 trucks from lowest to highest mileage.

Ranking by itself may be a useful technique. We now know that the truck with the greatest mileage is Truck 11 and the truck with the least mileage is Truck 7.

The median is the value at the midpoint of the ranks. The midpoint may be determined by inspecting the ranks or by adding the lowest rank number and the highest rank number and dividing by two. Thus rank 1 plus rank 11 divided by 2 equals rank 6.

The value at rank 6 is 13,000, giving a median of 13,000 miles. This number does not coincide with the mean of 16,000 miles. The mean and the median are only identical when the numbers in question are symmetrically distributed above and below the mean. In the event of an even number of units, the addition and division rule still applies, and the median becomes a theoretical halfway point. As an example, let us eliminate Truck 8 (50,000 miles). We now have 10 trucks and 10 ranks. When we add rank 1 plus rank 10 and divide by 2, we find a midpoint of rank 5.5. The theoretical value at rank 5.5 is 10,000 plus 13,000 divided by 2, or 11,500 as the midpoint. The median is thus 11,500 miles.

Table 2.3.   Recordable Injuries per Worker

Worker	Number of Injuries
John	1
Mary	2
Joan	0
Pat	0
Salman	3
Steve	4
Jim	2
Mike	0
Nick	1
Jorge	0
Total	13

2.5.3 Using the Median.   Although the mean is used more frequently than the median, the mean may give a deceptive representation if some of the values are extremely high or extremely low. In the case of the trucks, let us assume that Truck 8 had a high (but not impossible) mileage of 200,000 rather than 50,000 miles. The mean would then become 30,000 miles, which is an unrealistic representation of the fleet mileage. In this case the median, which is still 13,000, would be a better choice. In general, use the mean, but if some values are extremely high or extremely low, then use the median.

2.5.4 The Mean and Qualitative Categories.   The mean is best for quantitative categories found on a continuous scale, but may also be used for discrete qualitative frequencies. As an example, Table 2.3 shows the number of recordable injuries experienced by 10 workers in the last year. The mean is 1.3 (the total of 13 divided by 10) which is rounded to 1 injury as an average for the last year. Note that the workers with a zero value were included in the total.

2.6 Tabular and Graphic Presentation.   For control of numerical information and simplicity in communication, tables and diagrams are the best format.

2.6.1 Tables.   Large quantities and categories of numbers are most readily understood if placed in a table. Various types of numbers appearing in a table and numbers used in statistics are generally referred to as *data*. Data running up and down in a table are referred to as *columns,* and data running sideways are

Table 2.4. ABC Manufacturing Employee Demographics: Percentages of Total Workforce (1,000)

	Age			Sex		Seniority		Skill		High School Grad		Injury or Illness*	
	18–35	36–55	56+	M	F	1–10 Years	10+ Years	$10/Hour or Less	$10+/Hour	Yes	No	Lost Time	No Lost Time
Production worker	21	19	17	36	18	33	14	71	3	5	80	15	76
Nonproduction worker	12	18	13	14	32	15	38	10	16	14	1	2	7
Totals	33	37	30	50	50	48	52	81	19	19	81	17	83

Average age of all workers: 38 years

Average age of production workers: 34 years

Average age of nonproduction workers: 41 years

Average seniority of all workers: 9 years

Average seniority of production workers: 5 years

Average seniority of nonproduction workers: 12 years

Median wage of all workers: $ 9.00

Median wage of production workers: $ 8.00

Median wage of nonproduction workers: $11.00

*Total recordable illnesses and injuries for 1990: 105.

referred to as *rows*. The purpose and contents of the table should be clearly explained at the top. The columns and rows should have understandable headings. Tables should be given consecutive numbers if more than one is used in a report, article, or book. The text should refer the reader to the correct table in question. See Table 2.4 in the case study in section 2.7 for an example.

2.6.2 Pie Charts and Bar Graphs. Both pie charts and bar graphs are simple graphic representations showing proportionate numerical relationships. The pie chart is based upon a circle. The entire circle, or 360 degrees, represents 100% of a category. The area of a smaller segment is determined by multiplying 360 degrees by the percentage in question. Figures 2.1 and 2.2 in the following case study are pie charts. Bar charts are based upon percentages or frequencies usually along a vertical scale. If added up together, the several bars used in a bar graph should equal 100% or all of the members in the category. Figure 2.3 in the case study is a bar graph. Pie charts and bar graphs follow the same labeling rules as tables except that the title appears at the bottom rather than the top.

2.7 Case Study. Greg Farentino has just been hired as the Safety Officer of ABC Manufacturing, Inc. He wishes to find out about its 1,000 employees so that he can eventually do some effective safety programming. He also wants to include this information in a preliminary report to his superior. Greg has been able to acquire information by computer-aided enumeration from company records. He has decided to categorize the employees. Some of the more obvious categories he has chosen are age, sex, education, occupation, and injury and illness occurrences. Some less obvious categories are seniority and skill level.

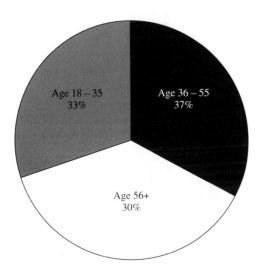

Figure 2.1. ABC Manufacturing: Age Distribution (Total Workforce of 1,000)

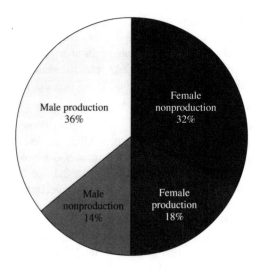

Figure 2.2.   ABC Manufacturing: Sex of Production and Nonproduction Workers (1,000)

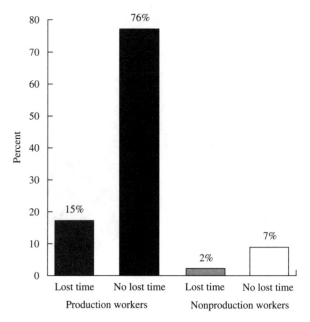

Figure 2.3.   ABC Manufacturing: Recorded Injuries and Illnesses, 1990 (105 Cases):
Lost Time versus No Lost Time for Production and Nonproduction Workers

The defined cutoff for high or low seniority is 10 years with the company. The defined cutoff for skill level is $10 an hour in wages. Greg can report the number of employees in each category, but he decides to use percentages, which he believes are more meaningful. From the data collected, he was also able to calculate the average age, average seniority, and the median hourly wage of various categories of workers. These will be reported along with the preceding data. All information appears in Table 2.4 and Figures 2.1, 2.2, and 2.3. Greg used pie charts for most of the information but decided that bar graphs would present other information more clearly. A good deal of judgment is involved in setting up tables and choosing the type of graphic representation. No two researchers will present the information in exactly the same way.

Problems

2.1. A company nurse has isolated the medical records of 48 employees who have been diagnosed with carpal tunnel syndrome. Table 2.5 lists the age and sex of these employees.
   a. From the entire list, determine the frequency and percentages of males and females in this population.

Table 2.5. Records Containing Carpal Tunnel Syndrome

Sex	Age	Sex	Age	Sex	Age	Sex	Age
M	32	F	51	F	48	M	26
F	49	M	32	M	28	F	55
F	53	F	58	F	42	M	25
F	49	M	31	F	58	F	59
F	60	F	48	F	50	M	33
M	24	F	55	M	26	F	51
F	45	M	31	M	31	F	61
M	27	F	45	M	20	F	53
F	46	M	30	F	54	F	52
F	52	F	59	M	28	F	49
F	28	F	48	F	49	F	61
F	40	F	39	F	50	F	53

b. From the entire list, determine the frequency and percentages of young, middle-aged, and older workers in this population (use 18–35 years = young, 36–55 years = middle-aged, and 56+ = older workers).

c. Enumerating the entire list, determine the mean and the median age and the age range of the population.

d. From the entire female list, determine the mean and the median age and the age range of females.

e. From the entire male list, determine the mean and the median age and the age range of males.

f. Draw a simple random sample of 27 from the entire list, and determine the mean and the median age and the age range of the sample.

g. Draw a stratified random sample of 27 from the entire list, and determine the mean and the median age and the age range of the sample.

h. Compare the results of the enumeration in (c) and the two sampling procedures in (f) and (g), and state some conclusions.

i. What general conclusions can you draw about this population of workers diagnosed with carpal tunnel syndrome?

j. Present the original information as well as information acquired in (a) to (e) in the form of tables, pie charts, and bar graphs.

2.2. The Vice President of Manufacturing of a large firm is concerned that manufacturing employees are not using assigned personal protective equipment (PPE). Inspectors have been assigned to spot-check various workstations to determine if PPE is being used by the worker at the workstation. The inspector is to report "Yes" or "No" on the use of PPE. While observing the worker, what other related information can the inspector report that might be useful to the VP?

2.3. A local hospital has recorded 120 industrial injuries that were treated in its emergency room. The parts of the anatomy affected and the frequency of occurrence are listed in Table 2.6.

a. Rank the frequencies.

b. Draw a pie chart and a bar graph describing these frequencies. Which diagram do you think communicates the information more effectively?

Table 2.6. Industrial Injuries Treated

Face, head	5	Leg(s)	2
Eye(s)	38	Other	55
Finger(s)	20		

    c. What is unsatisfactory about these categories, and what can be done about this in the future?

2.4. A company has noted that, for the prior year, its 95 recordable injuries from falls were distributed as follows:

Falls at same level: lost workday injuries = 21
                                 no lost workday injuries = 47

Falls from above: lost workday injuries = 20
                               no lost workday injuries = 7

    a. Place the frequencies and their percentages into a table.
    b. Draw a pie chart and bar graph describing the injuries.
    c. In this particular case, does the pie chart or the bar graph communicate more effectively?

Chapter 3

The Before With Control Design

3.1 Before Only Limitations. In the previous chapter, we discovered that the before only design is a highly flexible tool for descriptive research. We found that the members of a population can be placed into a large number of categories and that by calculating frequencies and percentages for the categories, we obtain quantifications that improve our understanding of the population. The use of numbers, however, exposes an important limitation of this research technique; that is, a number standing by itself has no special meaning unless it is compared to other number(s). To give an example, the Johnson Company carpentry shop has gone 125 working days (a full half-year) without a recordable injury or illness. Is this good or bad? Commendable or not commendable? An absolute opinion might be based upon personal philosophies, lifetime experiences, and a knowledge of the safety and health profession. A more desirable relative opinion might be based upon a numerical comparison of the various departments of the Johnson Company as shown in Table 3.1.

From the table, we can now see that the carpentry shop, whose record appeared good in an absolute sense, has a poor relative standing among the other shops. It must be remembered that numbers used for descriptions should, whenever possible, be provided with comparative numbers from other group(s).

3.1.1 The Second Design. The second research design we will be concerned with is known as *before with control*. This design always contains a *control population*, or *comparison group*, against which the population of concern, or *study group*, is compared. (see Figure 3.1). To establish a logical research design, some elemental rules must be followed in using and choosing a control group.

Table 3.1. Johnson Company Departments: Days without Injury or Illness

Rank	Department	Days without Injury or Illnesses
1	Motor pool	274
2	Assembly	260
3	Crating and packing	231
4	Metal stamping	203
5	Machine shop	175
6	Carpentry shop	125

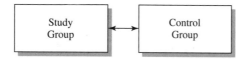

Figure 3.1. Before With Control Design

3.1.2 Criterion of Measurement. The two groups should be compared by using a single category or variable. This variable should be measured by the same criterion, and the criterion should be logically derived. For example, it would be incorrect to compare one group's injury and illness record against the other group's absentee rate or to measure one group by hours and minutes and the other group by cost containment. The comparison of the departments of Johnson Company in Table 3.1 is consistent in using days without injury or illness to measure each department. In future chapters, though, we may conclude that it would be more logical to compare the departments by the number of recordable injuries and illnesses per worker hours rather than by days without injury or illness.

3.1.2.1 Identical Characteristics. The two groups should be as identical as possible. The standard advice about not comparing apples and oranges applies to the study group versus the comparison group. The carpentry shop of Johnson Company is being compared with other departments whose personnel, processes, and equipment are not the same as those of the carpentry shop. Ideally, the carpentry shop should be compared with another, almost identical carpentry shop. This ideal, although highly desirable, cannot always be attained, and variations in characteristics between groups may be tolerated depending upon the judgment of the researcher and the research question at hand.

3.1.2.2 Identical Size. When comparing some variables, such as injury and illness occurrences, it is important to consider the relative sizes of the study group and the control group. In the case of Johnson Company, we may discover that the carpentry shop is the largest department in the corporation. If it is compared to the crating and packing department, which is a similar but smaller operation, then the carpentry shop's safety record may not be bad after all. The two following techniques may be used for equating the size factor among groups.

3.1.2.3 Comparisons. When studying injury and illness occurrences, the relative sizes between groups may be compared by (1) the number of people in each group or (2) the "busyness" of the group. When numbers of people are compared, only full-time members of one group should be compared with full-timers of the other group. To ensure a logical comparability, recent arrivals or part-timers in the groups should not be considered. The best way to compare "busyness" of the groups is to look at the number of working hours expended within a given period of time, particularly for similar categories of workers such as production workers in each group. It is important to be careful about overtime, which increases the hours reported by 1.5 or 2 but does not truly reflect the number of hours actually worked. Overtime hours should be properly modified for sake of comparability. Occasionally, a third method is used to compare sizes of groups. It involves comparing the number of units or objects found within each group such as machines, vehicles, square feet of floor space, and the like.

3.1.3 Ratios or Indexes. These comparison techniques can also be made more intelligible if a ratio is used. This is done by dividing the number of people or hours expended by the number of injuries and illnesses. Chapter 9 will describe some of the standardized indexes and ratios for injuries and illnesses established by ANSI (American National Standards Institute), BLS (Bureau of Labor Statistics), and OSHA (Occupational Safety and Health Administration).

3.1.4 Statistical Application. As already explained, the before with control technique requires a comparison between two groups. To make this comparison, a statistical tool must be implemented. The two following case studies will use an inferential statistical tool known as *chi square* (χ^2). By using this method, the researcher can determine whether a difference probably exists or probably does not exist between the two groups being observed. The discussion of χ^2 in this chapter is supplemented by an additional in-depth description in Appendix B.

3.2 Case Study. Smith is a safety-conscious individual who has just been hired as the supervisor of Department X. Department X is an electronics assembly operation that consists of a large number of workers putting together small electronic components while seated at workbenches. Smith is interested in the injury and illness record of his department and also wants to know how his department compares with others. Smith requests the aid of the Safety Officer, Davis. Davis tells Smith, who is new to the company, that there is another electronics assembly operation in the company, Department Y. Department Y's operations are very similar to Department X's except that Department Y has approximately twice as many workers as Department X. Davis suggests making a comparison with Department Y and provides Smith with the injury and illness records of both departments for the year 1996. Davis also helps Smith obtain from the payroll section information on the number of employees and hours worked in both departments. To attain a good design and logical comparability between the departments, Smith decided on the following criteria:

1. Only injury occurrences would be studied; illnesses would not be included in the study.
2. Only full-time workers from both departments would be included in the study. This corporation defines a full-time worker as someone who works at least 1,900 hours for at least 49 weeks during the year.
3. Injuries in overtime in both departments were not included in the study.
4. An injured worker was defined as a worker who experienced one or more recordable lost-time or no-lost-time injuries during the year. A non-injured worker was defined as a worker who experienced no recordable injuries during the year. Table 3.2 shows the data Smith acquired from the records.

The table indicates that in 1996 there were 185 full-time workers in both departments with 59 full-time workers in Department X and 126 in Department Y. Of these workers, 38 had at least one injury in the course of the year; 18 of the injured workers were in Department X and 20 were in Department Y. There were 147 workers who had no injuries; 41 of these workers were in Department X and 106 were in Department Y. The table implies that Department Y's injury record is better than Department X's because Department Y has twice as many

Table 3.2.   Noninjured and Injured Workers (1996)

	Number of Noninjured Full-Time Workers	Number of Injured Full-Time Workers	Total
Department X	41 (A)	18 (B)	59 (A + B)
Department Y	106 (C)	20 (D)	126 (C + D)
Total	147 (A + C)	38 (B + D)	185 (N)
AD = 41 × 20 = 820	BC = 18 × 106 = 1,908		

workers as Department X but far less than twice as many injuries. Implications, however, are not enough; a statistical test must be applied.

A χ^2 test will determine whether, at a probabilistic level of 95%, Smith's study shows a true difference between the injury occurrences of both departments or whether this study's outcome is simply a rare chance occurrence.

In the following formula, N represents the total number of workers in both departments, and A, B, C, and D represent the various values labeled in the table. The absolute notation $|AD - BC|$ indicates that after the computation is performed, the resulting $+$ or $-$ sign is eliminated. For greater simplicity the additions and multiplications needed $(A + B, C + D, A + C, B + D, AD, BC)$ were done in the table. The formula and its computation upon the data follow:

$$\chi^2 = \frac{N\left(|AD - BC| - \frac{N}{2}\right)^2}{(A + B)(C + D)(A + C)(B + D)}$$

$$\chi^2 = \frac{185\left(|820 - 1{,}908| - \frac{185}{2}\right)^2}{(59)(126)(147)(38)}$$

$$\chi^2 = \frac{185(1{,}088 - 92.5)^2}{41{,}526{,}324}$$

$$\chi^2 = \frac{183{,}338{,}746.3}{41{,}526{,}324} = 4.415$$

The final value of χ^2 is 4.415. This particular χ^2 design (two groups, Department X or Department Y, and two categories, injured or noninjured) requires a value of at least 3.84 in order to state a significant difference between the departments at a 95% level of probability. Since the χ^2 value arrived at (4.415) is greater than the minimum χ^2 value required (3.84), it may be stated with 95% confidence that Department X is different from Department Y in terms of the incidence of injuries. Stated in another manner, there is something about Department Y, safetywise, that is better than Department X. Note that if the χ^2 value were less than 3.84, the conclusion would be that there is no significant difference between the departments.

3.2.1 Case Continued. Smith is trying to find out what differences between Department X and Department Y might explain why Department Y has been performing in a superior fashion to Department X. He learns from Safety Officer Davis and from company records that the supervisor of Department Y has been faithfully sending almost all of her workers to quarterly safety training sessions given by the Safety Department. Smith also observes from records in his own department that his predecessor was not very strict about sending workers to safety training. He suspects that there is a difference in training patterns between

Table 3.3.   Safety Training of Full-Time Workers

	Number of Full-Time Workers Attending Safety Training	Number of Full-Time Workers Not Regularly Attending Safety Training	Total
Department X	53 (A)	6 (B)	59 (A + B)
Department Y	125 (C)	1 (D)	126 (C + D)
Total	178 (A + C)	7 (B + D)	185 (N)
AD = 53 × 1 = 53	BC = 6 × 125 = 750		

the departments. The breakdown of regular safety training by department is shown in Table 3.3.

Applying a χ^2 to this data, Smith obtains the following result:

$$\chi^2 = \frac{185\left(|53 - 750| - \dfrac{185}{2}\right)^2}{(59)(126)(178)(7)}$$

$$\chi^2 = \frac{185(697 - 92.5)^2}{9,262,764}$$

$$\chi^2 = \frac{67,602,746.25}{9,262,764} = 7.29$$

The χ^2 of 7.29 exceeds 3.84 and indicates that there is a difference in training patterns between the departments. A χ^2 of less than 3.84 would have meant that there is no difference between the departments' training patterns. This difference may not be the causal factor in the performance of Department X, however; instead it may be only one of several causal factors. Smith should certainly pay attention to this problem and should augment safety training, but he should also be on the lookout for other possible causes that he may identify by further research.

3.2.2   The Before With Control Design.   This design may be used to measure safety performances between groups and to compare characteristics between groups. As in this case study, safety performance can be measured by frequency. Severity may also be measured by forming categories such as Lost Time of Less Than 10 Days, Lost Time of More Than 10 Days, and the like. The method lends itself to enumeration or to sampling, which does not necessarily have to be very rigorous. The χ^2 technique that it uses is a conservative technique that does not always readily lead to significance, and it has a correction for smaller frequencies.

Problems

3.1 A company has four almost identical metal stamping shops, Shops 1, 2, 3, and 4. The Safety Officer has noted, during occasional visits, that the foreman of Shop 4 allows the machine operators to use the machines while the guards are not in place. The Safety Officer is not certain that the foreman is being excessively negligent and decides to gather information. The Safety Officer sends a safety inspector into each of the four shops and asks the inspector to choose 10 machines at random and note whether each machine is guarded or not guarded. The results are in Table 3.4.

   a. Execute a χ^2 comparison upon the data. Hint: Compare Shop 4 against a comparison group. The comparison group should consist of two numbers representing the totals of Shops 1, 2, and 3: unguarded machines equal 3, and guarded machines equal 27.

   b. Has the Shop 4 foreman been any more negligent than other foremen?

3.2 A Safety Officer believes that most of the company's severe injuries are occurring in the forging shop. He collects the data in Table 3.5, which compare injuries occurring in the forging shop with those occurring in

Table 3.4. Unguarded versus Guarded Machines

	Number of Unguarded Machines	Number of Guarded Machines
Shop 1	1	9
Shop 2	0	10
Shop 3	2	8
Shop 4	3	7

Table 3.5. Injuries per 10,000 Worker Hours (1996)

	Number of Injuries with No Lost Time	Number of Injuries with Three or More Days Lost Time
Forging Shop	5	25
All other manufacturing shops	17	10

all other manufacturing shops in the company in 1996. The Safety Officer has allowed for differences in size by collecting frequencies based upon 10,000 production worker hours.

a. Perform a χ^2 comparison based on the data in Table 3.5.

b. What is the conclusion?

3.3 A fire insurance underwriter is convinced that a sprinkler system is the best defense against total loss of a building from fire. By random sample she collects the data in Table 3.6 from her company's claims department for the year 1996. Does a sprinkler system make a difference in preventing a total loss?

Table 3.6. Losses from Fire (1996)

	Buildings Totally Lost	Buildings Partially Damaged
Sprinklers	23	115
Nonsprinklers	78	110

Chapter *4*

The Before and After Design

4.1 Review. In Chapters 2 and 3, we explained the before only design and the before with control design. We discussed the limitations of the before only design and the importance of making comparisons between groups, which led us to the use of the before with control design. These two simple techniques are very frequently used and belong within the scope of descriptive research as explained in section 1.3.3 in Chapter 1. Unfortunately, descriptive research is limited to observations at a single point in time. It does not necessarily take into consideration changes over time or the effects of any changes that management may have introduced into an organization.

4.2 Change over Time. We will now consider the *before and after design,* which was introduced in section 1.3.4 of Chapter 1 and is described in Figure 4.1. This design belongs within the area of experimental research and involves

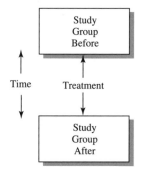

Figure 4.1. Before and After Design

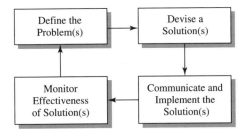

Figure 4.2. Scope and Functions of a Safety Professional

Source: Dan Petersen, *Techniques of Safety Management.*

change over time. This design is of great value to safety and health professionals because it helps them determine whether they are doing an effective job for their organization.

Figure 4.2 shows how the American Society of Safety Engineers defined the scope and functions of the safety and health professional in 1966 and again in 1994. Figure 4.2 is as appropriate now as it was in 1966. Stated in simpler terms, the functions of a safety and health professional are to (1) discover and define safety and health problem(s) within the organization, (2) devise solutions to the problem(s), (3) communicate and implement the solution(s) within the organization, and (4) monitor the effectiveness of the solutions. "Standing still is really nothing more than moving backward," so it is important for the safety and health professional to continuously seek beneficial changes in the organization by using the four steps in Figure 4.2. Note that the first step (defining the problem) readily coincides with the techniques of descriptive research described in Chapters 2 and 3, while the fourth step (monitor effectiveness) coincides with the technique of experimental research.

4.3 Before and After Procedure. As described in section 1.3.4, the before and after design involves the following steps:

1. Initially observe a group.
2. Effect a change within the group.
3. Observe the group after the change.

   Each activity will be discussed in turn.

4.3.1 Initial Observation. The procedures and rules for initially observing a group are the same as those described in Chapters 2 and 3. The characteristics that are observed must be logically and objectively chosen and must relate to the research question at hand. Observation results must be fairly, objectively, and consistently reported and recorded for future use. The observation is generally made upon one particular characteristic related to a problem area within the organization.

4.3.2 Effecting a Change. The process of effecting a change in a group may also be referred to as a *modification*, a *manipulation*, or, as in experimental research procedures, a *treatment*. The treatment may be of a physical nature such as guarding machines, grounding electrical equipment, or increasing illumination levels in a group's environment. The treatment may also be of a behavioral/cognitive nature such as training, motivating, or informing members of the group. Finally, the treatment may be of a managerial nature such as reprogramming, reorganizing a group, or rescheduling procedures within a group. The application of the treatment should be verifiable (are safety posters now hung in the plant?) and quantifiable (were 10, 20, 30, etc. posters hung in the plant?).

4.3.3 Secondary Observation. After the treatment, the group is reobserved. The same characteristic that was examined during the primary observation is observed again.

4.3.3.1 Prerequisites to Secondary Observation. It is important that the secondary observation (1) examines the same characteristic as the initial observation and (2) uses the same methods of analysis and observation as the initial observation. In the case of hanging safety posters, for example, if the initial observation was the frequency of injury and illness per 100 workers, then the secondary observation should also be the frequency of injury and illness per 100 workers.

4.4 Case Study. Roger is a member of the Safety Department of the West Coast plant of a large manufacturing company. Roger is currently working on his master's degree in ergonomics at a local university. Because of his interest in this field, Roger's boss, the safety director, has put Roger in charge of matters related to ergonomics. Roger regularly attends meetings of the West Coast Plant Safety Committee, which is composed of managers and representatives of labor. Recently, the committee has been concerned about the high incidence of low back injuries among employees. Some of the committee members, particularly those representing labor, are in favor of providing lumbar support back belts to workers who perform frequent manual materials handling. Roger is hesitant to endorse this idea because he has read that the effectiveness of back belts has not yet been proven, but he wishes to keep an open mind. He also does not want to give the representatives of labor the impression that the company is not interested in their welfare. Roger suggests the following experimental procedure, whose implementation is supported by both upper management and the Safety Committee:

1. During 1995, Roger will determine which groups of workers are currently most exposed to low back injury.
2. The number of newly reported low back injuries within these groups during 1995 will be counted.

3. The workers within these groups will be provided with back belts and will be trained in their correct use.
4. These workers will use the belts for one whole year (1996).
5. In early 1997, the number of newly reported low back injuries within these groups during 1996 will be counted.
6. The number of incidents for 1995 will be compared statistically with the number of incidents in 1996 to determine whether the treatment (use of back belts) has been effective.

Roger begins to execute the research design. By searching the records, he has discovered that the employees working on the loading platform and the truck drivers have the greatest incidence of low back injury. He therefore includes only these workers in the study. He proceeds as described above with the intention of eventually comparing 1995 injuries against 1996 injuries. Roger wishes to maintain as strict a comparability between the two years as possible and thus intends to take relative "busyness" of the two years into consideration. Roger discovers the data presented in Tables 4.1, 4.2, and 4.3.

Table 4.1.   Low Back Injuries

	Loading Platform	Motor Pool	Total
1995	9	6	15
1996	5	7	12

Table 4.2.   Payroll Hours

	Straight Time		Overtime		
	Loading Platform	Motor Pool	Loading Platform	Motor Pool	Total Actual Hours
1995	60,000	120,000	40,000	0	200,000
1996	60,000	130,000	0	0	190,000

Hours-to-employee conversion:

1995: 200,000 hours/2,000 hours = 100 employees

1996: 190,000 hours/2,000 hours = 95 employees

Table 4.3.   Injured and Noninjured Employees

	Employees Injured	Employees Not Injured	Total
1995	15 (A)	85 (B)	100 (A + B)
1996	12 (C)	83 (D)	95 (C + D)
Total	27 (A + C)	168 (B + D)	195 (N)

$AD = 15 \times 83 = 1{,}245$ $BC = 85 \times 12 = 1{,}020$

These tables combine before and after information on the number of back injuries and the payroll hours of employees for both the loading platform and the motor pool. Note that Roger made an adjustment for overtime (which in this company is double time) in Table 4.2. Roger was not able to readily determine the number of employees in the before and after groups. He was, however, able to work up an equivalence between hours worked and number of employees by dividing the number of hours worked by 2,000 (the average number of hours worked by an employee in this company). Table 4.3 presents the information needed to perform the χ^2 test. Solving for χ^2 in the same manner as in Chapter 3 gives a χ^2 value of only .074, which is less than the necessary 3.84. This indicates that there is no significant difference between the before and after groups, and thus Roger concludes that the use of back belts had no effect upon the workers. The details of the solution follow:

$$\chi^2 = \frac{N\left(|AD - BC| - \dfrac{N}{2}\right)^2}{(A + B)(C + D)(A + C)(B + D)}$$

$$\chi^2 = \frac{195\left(|1{,}245 - 1{,}020| - \dfrac{195}{2}\right)^2}{(100)(95)(27)(168)}$$

$$\chi^2 = \frac{195(225 - 97.5)^2}{43{,}092{,}000}$$

$$\chi^2 = \frac{195(127.5)^2}{43{,}092{,}000}$$

$$\chi^2 = \frac{3{,}169{,}968.75}{43{,}092{,}000} = .074$$

Problems

4.1 Ms. Yoshida is an industrial hygienist working for a large insurance company. The company has moved its main headquarters from an old office building to a newly erected office building. The new building is now occupied by approximately the same 4,500 employees who were working in the old building. After one year the Human Resources Director notes that absenteeism due to respiratory complaints has increased sharply. The Human Resources Director has read about "sick building syndrome" and is concerned that the new building has been causing the increased respiratory absenteeism. Ms. Yoshida is asked to investigate. She searches through attendance records and counts the number of employees who have called in sick one or more times, citing respiratory complaints, in the last two years. She discovers the following:

   Year 1995 (old building): 259 employees
   Year 1996 (new building): 562 employees

   a. Determine whether there is any significant difference between the frequency of respiratory illness in the new and old buildings.
   b. If there is a difference, what does this imply?

4.2 Mr. Dietrich is a member of a corporate training department. His responsibility is safety training. In the past, only one hour has been dedicated to a safety orientation for new employees. Mr. Dietrich believes that this is inadequate and that the orientation should be extended to four hours. He tells his boss that extending the orientation will substantially reduce the number of accidents occurring to employees during their first six months on the job. His boss tentatively allows an increase in orientation time provided that it can be proved to be effective. Mr. Dietrich (1) immediately increases the new employee safety orientation from one to four hours, (2) looks into personnel and injury records for the last 12 months and counts the total number of employees with six or fewer months on the job, (3) notes how many of these new employees experienced at least one reportable injury, and (4) waits one year and repeats this data collection procedure. His findings are summarized in Table 4.4. Using these data, determine whether the increase in safety orientation time has had a beneficial effect.

4.3 Joe, the company industrial hygienist, is convinced that lubricating the presses in the printing plant more frequently will substantially lower the objectionable sound levels in the plant. Sound level readings vary

Table 4.4. Safety Orientation Time and Injuries

	New Employees Injured	New Employees Not Injured
Before (1-hour orientation)	21	92
After (4-hour orientation)	9	101

Table 4.5. Noise Levels and Stepped-up Lubrication

	Number of Squares at 85 dba or More	Number of Squares at Less Than 85 dba
Before step-up	6	19
After step-up	4	21

from area to area and from time to time within the plant. To test his theory, Joe takes the following steps:

- He divides the plant's floor area into an imaginary 25-square checkerboard.
- At what he considers a peak activity time, he takes a sound level reading at each square and notes whether the reading was 85 dba or above or below 85 dba.
- He then asks the maintenance department to step up lubrication frequency.
- After an appropriate wait, Joe repeats the testing procedure at a peak activity time. Table 4.5 presents his findings. Did increasing the frequency of lubrication make any statistical difference?

4.4 Mr. Johnson is the risk manager of a large corporation. Traditionally, the corporation has been self-insured in worker's compensation insurance. To control worker's compensation losses Mr. Johnson has been using the services of a local consulting firm. He has recently decided to forgo the services of the consulting firm and organize his own loss control department under the supervision of an assistant risk manager. Mr. Johnson decides to find out if this was a good decision. He (1) counts the number of employees having at least one reportable injury in the 12 months prior to organizing the loss control department,

(2) counts the number of hours worked by production employees for the same period, (3) waits 12 months during which time the loss control department is operational, and (4) performs the same data collection procedure as before. Table 4.6 presents his findings. Has the new loss control department been effective?

Table 4.6. Loss Control Department versus Consulting Firm

	Number of Employees Injured	Production Hours
Consulting firm	55	2,158,623
Loss control department	20	1,945,920

Note: Production workers work full-time and put in 2,000 hours a year.

Chapter *5*

The Before and After With Control Design

5.1 The Final Design. The final design for the *before and after with control technique* is a combination of the before and after technique and the before with comparison technique. It is an experimental procedure similar to the before and after technique, but it is superior because it includes a control in the form of a comparison group that provides stronger support for the researcher's final conclusions. This technique is frequently used in educational, psychological, and social research. Figure 5.1 describes this technique.

5.2 Explanation. The before and after with control technique takes into consideration the possibility that the group being studied and receiving the treatment in question would have changed whether the treatment had been applied or not. This change might have been caused by various internal or external forces. For example, an exercise physiologist believes that isometric exercises are beneficial to people who are inclined to be sedentary. To test his theory, he gives strength tests to a group of sedentary early adolescents. He then asks the

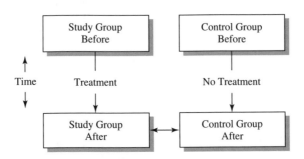

Figure 5.1. Before and After With Control Design

adolescents to perform isometric exercises regularly for six months. After the six-month period, he readministers the same strength tests. He discovers that the adolescents are substantially stronger and declares that isometric exercise is beneficial for strengthening sedentary people. He reports his conclusion to one of his colleagues, who listens, thinks, and then asks: "Since these are growing adolescent children, how do we know they wouldn't have gotten stronger, anyway, without the isometric exercise?" The exercise physiologist had not considered the possibility of an internal change in the adolescents in the form of maturation. His research would have been better if he had used a comparison group as well as an experimental group. The next section describes a more appropriate procedure that takes this type of unexpected (or unsuspected) change into account.

5.3 Before and After With Control Procedure. This procedure involves the following steps:

1. Choose two groups. If at all possible, the two groups should be similar in size and important characteristics.
2. Test or observe both groups with identical testing or observation procedures.
3. Apply a treatment to one group (the experimental group) but not to the other group (control group).
4. Wait an appropriate amount of time.
5. Retest or reobserve both groups using the same test or observation procedure that was used initially.
6. For both groups, compare the results of the original test or observation with those from the retest or reobservation.
7. If there is a statistically significant change in the experimental group but not in the control group, then the researcher's assumptions are correct and the treatment has had an effect.

5.4 Case Study. Ms. Weiss is the safety officer of a medium-sized corporation. The safety record for her company is above average. Nevertheless, Ms. Weiss is concerned because many employees are injured during off-hours in their homes, on the highways, or in public places. She therefore decides to design and implement an Off-the-Job Safety Program. She is not totally certain that such a program will succeed, however, and she is reluctant to try it for the entire organization. The company is divided into two plants, Plant 1 on the west side of town and Plant 2 on the east side. The plants are approximately the same size and have the same operations and the same types of employees. Ms. Weiss decides to try an Off-the-Job Safety Program in Plant 1 only. She does the following:

1. For the year 1995, she counts the number of full-time employees at each plant and also counts the number of full-time employees experiencing at least one off-the-job injury at each plant.

2. At the beginning of 1996, she introduces an Off-the-Job Safety Program in Plant 1 but not in Plant 2.
3. At the end of 1996, she collects the same type of data for each plant as she did in 1995.
4. Her collected data appear in Table 5.1.

In order to perform a χ^2 test similar to those in Chapters 3 and 4, Ms. Weiss changes the data as in Table 5.2. The first step is to perform a χ^2 test on Plant 2 (control group), using the data in Table 5.3:

$$\chi^2 = \frac{N\left(|AD - BC| - \frac{N}{2}\right)^2}{(A + B)(C + D)(A + C)(B + D)}$$

$$\chi^2 = \frac{987\left(|14{,}167 - 15{,}378| - \frac{987}{2}\right)^2}{(490)(497)(923)(64)}$$

$$\chi^2 = .035$$

Table 5.1. Off-the-Job Injuries to Full-Time Workers

	Plant 1		*Plant 2*	
	Total Full-Time Workers	**Total Full-Time Workers Injured**	**Total Full-Time Workers**	**Total Full-Time Workers Injured**
1995	506	32	490	33
1996	510	16	497	31

Table 5.2. Noninjured versus Injured Workers

	Plant 1			*Plant 2*		
	Noninjured Workers	**Injured Workers**	**Total**	**Noninjured Workers**	**Injured Workers**	**Total**
1995	474	32	506	457	33	490
1996	494	16	510	466	31	497

The next step is to compare, using χ^2, the before and after change of Plant 1 (experimental group). Using the data in Table 5.4, the calculation is as follows:

$$\chi^2 = \frac{1,016\left(|7,584 - 15,808| - \frac{1,016}{2}\right)^2}{(506)(510)(968)(48)}$$

$$\chi^2 = 5.05$$

The results of the χ^2 tests indicate a significant before and after difference for Plant 1 but not a significant before and after difference for Plant 2. Since the experimental group changed but the control group did not change, Ms. Weiss concludes that the Off-the-Job-Safety Program is an effective instrument that will provide beneficial results if applied under future circumstances.

5.5 Overview. The designs presented in Chapters 2, 3, 4, and 5 have progressed from simple to more complex and from minimally expensive and time-consuming to maximally expensive and time-consuming. The more complex, expensive, and time-consuming the design, the more logical, conclusive,

Table 5.3. Noninjured versus Injured Workers at Plant 2

	Noninjured Workers	Injured Workers	Total
1995	457 (A)	33 (B)	490 (A + B)
1996	466 (C)	31 (D)	497 (C + D)
Total	923 (A + C)	64 (B + D)	987 (N)

$AD = 457 \times 31 = 14,167$ $BC = 33 \times 466 = 15,378$

Table 5.4. Noninjured versus Injured Workers at Plant 1

	Noninjured Workers	Injured Workers	Total
1995	474 (A)	32 (B)	506 (A + B)
1996	494 (C)	16 (D)	510 (C + D)
Total	968 (A + C)	48 (B + D)	1,016 (N)

$AD = 474 \times 16 = 7,584$ $BC = 32 \times 494 = 15,808$

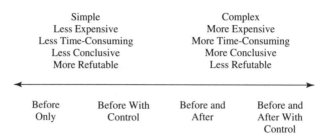

Figure 5.2.   Research Design Comparison

and irrefutable will be the results. Nevertheless, the less complex, less expensive, and less time-consuming studies should not be ignored. The choice of a design depends upon practical circumstances, the severity of the problem at hand, and the good judgment of the researcher. Figure 5.2 illustrates the relationship between complex and simple designs.

Problems

5.1 Ms. Marshal is the director of nurses at a prominent hospital. The hospital is experiencing the same problem as many other hospitals: low back injuries to nurses and nurse's aides caused by lifting and manipulating bedridden patients. Ms. Marshal has learned that a mechanical lifting device to aid in moving patients is available, but she is reluctant to ask the hospital administration to make an expensive investment in this device until she is sure that it will substantially diminish low back injuries. She has permission to make some limited purchases and do some fact-finding, so she does the following:

     a. For a prior 12-month period, she counts the number of employees experiencing a low back injury for the first time in Wing A of the hospital and also counts the number of full-time nurses and nurse's aides working in this wing.

     b. She does the same for Wing B of this hospital after satisfying herself that Wing B has approximately the same characteristics as Wing A.

     c. She then purchases and provides the lifting devices for Wing A and gives adequate instructions in their use.

     d. She does not purchase any of the devices for Wing B.

     e. After 12 months in which the devices were used in Wing A, Ms. Marshal performs the same data collection procedures as in steps (a) and (b) above.

Table 5.5. Injuries Before and After Use of the Lifting Device

	Wing A		*Wing B*	
	Number of Injured Employees	Total Number of Employees	Number of Injured Employees	Total Number of Employees
Before	20	82	18	100
After	16	78	19	105

Table 5.6. Injuries With and Without a Safety Bonus

	West Coast		*East Coast*	
	Total Hours Worked	Number of Injured Employees	Total Hours Worked	Number of Injured Employees
1995	816,000	48	201,000	12
1996	792,000	43	205,000	4

Note: A typical employee works 2,000 hours a year.

Her findings appear in Table 5.5. What are her conclusions about the devices?

5.2 Mr. Flaherty is the CEO of a prominent corporation. His safety officer has made an important suggestion that might improve the company's safety performance. He has suggested a Safety Bonus Plan that will award a safety bonus every month to each employee who has not experienced an injury. The plan also dictates that the bonus will be temporarily lost if the employee is injured. Considering the cost of this plan and its questionable success, Mr. Flaherty tentatively implements the plan in the following manner:

a. For the year 1995, he counts the total number of hours worked by all of his employees in the main plant on the West Coast. He also counts the number of employees in this plant who experienced at least one injury.

b. For 1995 he does the same for the smaller East Coast plant.

c. At the beginning of 1996, he implements the Safety Bonus Plan in the smaller East Coast plant only.

d. At the end of 1996, he performs the same data collection procedures as in steps (a) and (b) above.

Mr. Flaherty's findings appear in Table 5.6. Would a Safety Bonus Plan be a worthwhile endeavor for the entire corporation?

5.3 John is an industrial hygienist working for a large steel foundry. John would like to provide protective reflective clothing for employees working in hot areas. So far no severe cases of heat stress have been reported at the foundry, but John still believes that the protective clothing is essential. He decides to convince management of the need for this clothing by way of their pocketbook. He can do this by studying worker productivity. The plant has two major, reasonably identical, foundry operations, Line A and Line B. Each line is run autonomously and has its own foundry production workers and its own administrative and management personnel. John does the following:

a. For 1995, for each line, he collects production hours of foundry production workers and administrative and management hours for administrative and management workers. He also collects production totals in the form of tons of steel produced by each line.
b. For each line, he then determines the hours per ton of steel for production workers and the hours per ton of steel for administrative workers and managers.
c. He next issues reflective protective clothing to Line A production workers but not to Line B production workers.
d. At the end of 1996, John performs the same data collection procedures as in steps (a) and (b) above.

John's findings appear in Table 5.7. Has the protective clothing improved productivity?

Table 5.7. Hours per Tons of Steel for Foundry Workers

	Line A		Line B	
	Production Workers	Administrative and Management Personnel	Production Workers	Administrative and Management Personnel
1995	50	500	52	475
1996	10	505	53	488

Chapter 6

Correlations

6.1 Introduction. An important technique that may be of value to a researcher in safety and health is correlation. *Correlations* are a method of determining how strong the relationship is between two or more characteristics, which may also be called *variables*. The variables are usually measured by uniform large or small discrete increments such as years of experience, scores on an examination, readings on a measuring instrument, and the like. These increments are usually referred to as a *scale*. A correlation is expressed as a number from 0 to 1. The closer the number is to 1, the stronger is the direct or indirect relationship between the variables. The closer to 0, the weaker is the relationship. It is generally agreed that a correlation of .1 to .3 is a weak correlation, a correlation of .4 to .6 is a moderate correlation, and a correlation of .7 to .9 is a strong correlation. A correlation of 1 (which is unlikely to occur) is a perfect correlation in which there is perfect predictability between the variables. A correlation of 0 means no relationship at all. Correlations may be positive and marked with a plus sign or negative and marked with a minus sign. In a positive correlation, the measurements of both variables move in the same direction. A negative correlation occurs when the measurements move in opposite directions. As an example, a positive correlation is likely to occur between the number of years of education of a group of adults and the size of their annual income. An example of a negative correlation might be the age of a group of adults and their grip strength in pounds as measured on a dynamometer.

6.2 Procedure. The statistical procedure we will use is a *rank order correlation,* which relies on making comparisons between two variables based upon their relative ranked positions. Like the chi square (χ^2) technique used in

Table 6.1.   Injuries and Supervisor's Experience

Supervisor	Years of Experience	Number of Injuries per 100,000 Hours
Lisa	1	9
Harry	3	7
John	3	6
Ramon	6	4
Sam	7	3
Kevin	8	2
Mike	11	0
Jim	5	1
Rudy	2	5
Charlie	4	8

Chapters 3, 4, and 5, this procedure is a nonparametric technique that allows ample latitude in research designs and substantial freedom in the choice of groups and individual subjects. This procedure can only be used for two variables. Other correlational techniques may study more than two variables, but they are beyond the scope of this book.

6.3 Case Study. Bob is a safety officer of a corporation. He believes that the supervisor of a department is the key to a successful safety program. He is not certain, but observations have suggested to him that the departments with the best safety records are run by the most experienced supervisors in the corporation. By collecting data on the 10 supervisors who are in charge of various production activities, Bob is able to determine the years of experience they have had as supervisors in the corporation. He is also able to determine the number of injuries per 100,000 worker hours each supervisor's department experienced in the prior year. Table 6.1 shows these data.

Bob then ranks the supervisors by their experience from high to low and calls this column X. Next to each supervisor's experience rank, Bob places the injury experience rank from lowest to highest and calls it column Y. Mike, for example, ranked first in years of experience (11) and first in injury experience (0). Note that Harry and John have the same number of years of experience and thus were given identical rankings. The rearrangement appears in Table 6.2. Note also that the table contains two additional columns: d, which is the difference between columns X and Y, and d^2, which is the square of the value in column d. The computational formula will use the total of column d^2 and N,

Table 6.2. Ranking by Experience and Injury Frequency

	X Experience Rank	Y Injury Rank	d X − Y	d²
Mike	1	1	0	0
Kevin	2	3	−1	1
Sam	3	4	−1	1
Luis	4	5	−1	1
Jim	5	2	+3	9
Charlie	6	9	−3	9
John	7.5	7	+.5	.25
Harry	7.5	8	−.5	.25
Rudy	9	6	+3	9
Lisa	10	10	0	0
				30.50

$N = 10$

$d^2 = 30.50$

which is the number of members (supervisors) in this sample. Using these numbers, we can compute the correlation between a supervisor's years of experience and low injury frequency in the department as follows:

$$r = 1 - \frac{6 \sum d_i^2}{n^3 - n}$$

Spearman

$$\rho = 1 - \frac{6(\sum d^2)*}{N(N^2 - 1)}$$

$$\rho = 1 - \frac{6(30.50)}{10(10^2 - 1)}$$

$$\rho = 1 - \frac{183}{10(99)}$$

$$\rho = 1 - .185$$

$$\rho = .82$$

---

*Σ means "sum of."

P.3. p.165

The correlation is a substantial +.82. This is somewhat higher than is usually found in management research. One additional step is now needed. Table 3 (*p* values) at the rear of this book will indicate whether this is a significant correlation. Looking down the *N* column until we reach our *N* of 10, we see that at a 5% level of significance, a correlation of at least .648 is needed. The correlation of .82 exceeds this; thus, the correlation is significant. If the correlation was less than .648, it would have to be disregarded.

6.4 Discussion. This study shows a high positive correlation between years of experience and low injury frequency and has given Bob an important piece of information. He may now assume that experienced supervisors tend to run safe departments. He may thus encourage the management of his company to place experienced supervisors in departments where safety problems are present.

6.5 Caution. An important warning must be made about correlational studies: correlation does not imply causation. The supervisors' years of experience did not necessarily cause their good safety records. Their years of experience merely coincide with their good safety records.

Problems

 6.1 Mike is an industrial engineer and human factors specialist who is interested in how workers' error rates relate to production speed. His concern is that excessively rapid production leads to expensive rejects and at times to accidental trauma to the worker. The workers in one of his company's plants are working on piece rate and are thus motivated to work rapidly. He has timed a particular operation as averaging between 20 and 40 seconds depending upon the speed of the particular worker. He has also been able to count the number of average rejects (production errors) for each worker. Table 6.3 summarizes his findings.

  a. Compute a rank order correlation.
  b. Is the final result a negative or positive correlation?
  c. Is this a significant correlation?
  d. Is this a weak, moderate, or strong correlation?

 6.2 A safety director is interested in the consistency of the safety performance of her company's manufacturing departments. She believes that a department with a good safety appraisal will continue to have good appraisals into the future and a department with a bad appraisal will continue to have undesirable appraisals into the future. Semiannually, the safety department rates each department from bad (0) to excellent (100) by evaluating injuries, safety procedures, and safety training. Recent results are shown in Table 6.4.

Table 6.3.   Worker Speed and Average Rejects

Worker	Average Speed (Seconds)	Average Daily Rejects
1	20	15
2	38	6
3	37	5
4	29	12
5	22	13
6	25	14
7	40	5
8	32	10
9	26	14
10	39	8
11	21	12
12	24	11
13	30	7
14	31	4
15	32	9

Table 6.4.   Safety Ratings of Departments

Department	January Appraisal	July Appraisal
A	94	90
B	75	77
C	83	90
D	76	81
E	88	88
F	69	60
G	90	90
H	73	74
I	85	86
J	80	78

a. Compute a rank order correlation on the data.
b. Is the correlation significant?
c. Are the departments consistent in their safety performance?

6.3 An electronics assembly department contains 20 workbenches. Each workbench is continually occupied by the same full-time assembly worker. Benchtop illuminations vary from workbench to workbench and may be as low as 30 footcandles and as high as 45 footcandles. The company safety officer has observed that workers in this department have experienced a large number of cut and bruised fingers. He believes that illumination levels may have something to do with the frequency of injury. He is able to count the number of injuries at each bench over a year. Table 6.5 presents his findings. Is there any relationship between illumination level and injury?

Table 6.5. Injuries and Illumination Levels

Bench	Illumination Level	Number of Injuries	Bench	Illumination Level	Number of Injuries
1	38	7	11	36	7
2	40	0	12	33	10
3	31	8	13	43	1
4	45	8	14	40	10
5	44	4	15	40	6
6	40	6	16	38	3
7	32	2	17	34	0
8	33	9	18	35	6
9	39	1	19	42	2
10	39	5	20	43	9

Chapter *7*

Comparing Test Results

7.1 Discussion. In the course of research, groups of individuals or objects may be analyzed for their present or future characteristics. These characteristics may also be used as a basis of comparison against those of other groups. One frequently used method of group analysis is known as testing. *Testing* may consist of observations of individual performances of group members, or more traditionally, it may consist of individual responses on paper with pencil, better known as paper and pencil tests. Eventually, the overall test results of one group must be compared with those of one or more other groups. This chapter will present a procedure for comparing the test results of two or more groups. This procedure is known as the *Kruskal-Wallis one-way analysis of variance by ranks* and is also called the *H* test. Like the procedures described in Chapters 3, 4, and 5, it is a nonparametric technique. There is, however, an important prerequisite for the use of this technique. The test should be designed so that (1) the minimum score is zero and (2) there are equal increasing intervals between the zero score and the maximum score. Examples of tests that meet this requirement are a paper and pencil test whose scores range from 0% to 100%; measuring people's height in inches, weight in pounds, or age in years; and counting the number of errors committed by a machine operator during a specified period of time.

7.2 Case Study. Professor Enstrom is a member of the faculty of a Safety and Health Department of a local university. He teaches a course known as Principles of Hazard Control to matriculated undergraduate students, and he also teaches this identical course to local certificate students who have hands-on experience in industry. The professor wishes to compare the relative competence of the two groups of students. He decides to administer an identical final exam to both groups. The exam is composed of 100 questions worth 1 point each, and scores can run between 0% and 100%. The results appear in Table 7.1. Note that the groups are not equal in size. This situation is acceptable because this procedure is designed for groups of unequal as well as equal size.

Table 7.1.   Final Exam Grades: Two Groups

Certificate	Undergraduate
89	85
85	72
86	79
92	85
90	90
84	76
81	83
70	71
93	69
94	

Table 7.2.   All Scores Ranked

Score	Rank	Group	Score	Rank	Group
94	19	C	83	8	U
93	18	C	81	7	C
92	17	C	79	6	U
90	15.5	C	76	5	U
90	15.5	U	72	4	U
89	14	C	71	3	U
86	13	C	70	2	C
85	11	C	69	1	U
85	11	U			
85	11	U			
84	9	C			

Professor Enstrom discovers that the rounded average scores of the groups are 86% for certificate students and 77% for undergraduates. He is tempted to conclude that the certificate students are more competent but wishes to substantiate this conclusion with an *H* test.

His first step is to combine all 19 scores, regardless of group, and list them according to rank from the highest to the lowest as shown in Table 7.2. Note that the highest score has the highest rank and the lowest score has the lowest rank and that identical scores are given identical ranks according to the rules of ranking in Chapter 2.

The professor next creates Table 7.3 where the grades and ranks of the two groups are separated. This table contains several important symbols: *N* is the total number of scores in the study (19); *n* is the total number of scores in each group (10 and 9); *R* is the sum of the rank numbers for each group (125.5 and 64.5). Next the following formula is applied. Note that the numbers 12 and 3 are constant numbers that appear in all cases regardless of the type and size of the study.

Table 7.3.   Grades and Ranks of Each Group

Group 1 *Certificate* *n = 10*		*Group 2* *Undergraduate* *n = 9*	
Score	**Rank**	**Score**	**Rank**
94	19	90	15.5
93	18	85	11
92	17	85	11
90	15.5	83	8
89	14	79	6
86	13	76	5
85	11	72	4
84	9	71	3
81	7	69	1
70	2		64.5 = *R*
	125.5 = *R*		

$N = 10 + 9 = 19$

$$H = \left(\frac{12}{N(N+1)}\right)\left(\Sigma\frac{R_1^2}{n_1} + \frac{R_2^2}{n_2}\text{ etc.}\right) - 3(N+1)$$

$$H = \left(\frac{12}{19(19+1)}\right)\left(\frac{125.5^2}{10} + \frac{64.5^2}{9}\right) - 3(19+1)$$

$$H = \left(\frac{12}{380}\right)\left(\frac{15,750.25}{10} + \frac{4,160.25}{9}\right) - 3(20)$$

$$H = .032(1,575.03 + 462.25) - 60$$

$$H = .032(2,037.28) - 60$$

$$H = 5.19$$

The final result is $H = 5.19$. Although this result is usually satisfactory, it is suggested that a correction be made to compensate for any tying (identical) numbers that appear in the study, especially if a substantial number of them occur. In this study, the number 85 appears three times, and the number 90 appears two times. We assign a value of $t_1 = 3$ to the triple tie, and $t_2 = 2$ to the double tie and use the following formula for our correction:

$$H_c = \frac{H}{1 - \dfrac{\Sigma(t_1^3 - t_1) + (t_2^3 - t_2)\text{ etc.}}{N^3 - N}}$$

$$H_c = \frac{5.19}{1 - \dfrac{(3^3 - 3) + (2^3 - 2)}{19^3 - 19}}$$

$$H_c = \frac{5.19}{1 - \dfrac{24 + 6}{6,840}}$$

$$H_c = \frac{5.19}{1 - .004}$$

$$H_c = 5.21$$

The final corrected result is $H = 5.21$, which is only slightly more than the original $H = 5.19$. The next step is to determine whether this number (5.21) reaches a statistical significance. The Chi Square Table (Table 4) in Appendix D, which was used in Chapters 3, 4, and 5, is also used for this test. To use the table, we must first determine the number of degrees of freedom involved in this study. Degrees of freedom (*df*) are equal to the number of groups minus 1. In this study there are 2 groups minus 1, which gives a *df* of 1. Since the .05 level is usually required for significance, we search the table for 1 *df* and 5% and discover the required level is 3.84. Since 5.21 is greater than 3.84, we conclude that we have significance and that there is a difference between the certificate students and the undergraduate students. At this point it is important to note that the difference between the groups is based upon an exam devised by the professor. Professor Enstrom must ask himself whether the exam is a valid, competently designed exam that covers all areas of the subject and measures what it is supposed to be measuring. Books are available that may be of assistance in designing such a competent examination. Some suggestions about test design also appear in Chapter 14 of this book.

7.3 Another Example. This technique may also be used for comparing more than two groups and may be applied in two different ways: (1) many individuals may be tested only once or just a few times, or (2) only a few individuals may be tested many times. The prior example described a study with only two groups in which a fairly large number of people were tested only once. This example will have three groups, each composed of only one individual and that individual will be tested several times. Jody McKay is a Driver's Ed teacher in a local high school. She wishes to emphasize to her students the dangers of alcohol impairment and loss of driving skills. With the permission of her superior, who has in turn gotten the approval of the school's legal department, Jody sets up an experiment. She lays out a long, narrow, winding, driving lane demarcated by traffic cones on the school parking lot. Three student volunteers are asked to drive down the lane five times at a uniform speed of 20 mph. One of the students voluntarily drank a can of beer 15 minutes before driving the lane. The number of errors in the form of cones knocked over while driving the lane, were recorded for each driver. The results appear in Table 7.4.

Table 7.4.   Number of Errors for Each of Five Runs

Driver 1 (Alcohol)	Driver 2 (No Alcohol)	Driver 3 (No Alcohol)
6	2	1
8	0	3
7	3	3
5	4	3
9	3	2

Table 7.5. All Scores Ranked

Score	Rank	Group		Score	Rank	Group
9	15	1		3	7	3
8	14	1		3	7	3
7	13	1		3	7	3
6	12	1		2	3.5	2
5	11	1		2	3.5	3
4	10	2		1	2	3
3	7	2		0	1	2
3	7	2				

The rounded average error for each driver is 7, 2, and 2. We strongly suspect that the alcohol drinker is different from the other drivers, but we still prefer to submit this to a statistical test. As in the previous example, scores are lumped together and ranked, along with a notation as to which group (driver) the score belongs (see Table 7.5). Table 7.6 then presents the scores and ranks for each group. The H value is computed as follows:

$$H = \left(\frac{12}{N(N+1)}\right)\left(\Sigma\frac{R_1^2}{n_1} + \frac{R_2^2}{n_2} \text{ etc.}\right) - 3(N+1)$$

$$H = \left(\frac{12}{15(16)}\right)\left(\frac{65^2}{5} + \frac{28.5^2}{5} + \frac{26.5^2}{5}\right) - 3(16)$$

$$H = .05(845 + 162.5 + 140.5) - 48$$

$$H = 9.4$$

$$H_c = \frac{H}{1 - \dfrac{\Sigma(t_1^3 - t_1) + (t_2^3 - t_2) \text{ etc.}}{N^3 - N}}$$

$$H_c = \frac{9.4}{1 - \dfrac{(5^3 - 5) + (2^3 - 2)}{15^3 - 15}}$$

$$H_c = \frac{9.4}{1 - \dfrac{120 + 6}{3,360}} = \frac{9.4}{1 - .038} = 9.8$$

Table 7.6. Scores and Ranks for Each Group (Driver)

Driver 1 (Alcohol) n = 5		Driver 2 (No Alcohol) n = 5		Driver 3 (No Alcohol) n = 5	
Score	**Rank**	**Score**	**Rank**	**Score**	**Rank**
9	15	4	10	3	7
8	14	3	7	3	7
7	13	3	7	3	7
6	12	2	3.5	2	3.5
5	11	0	1	1	2
	$65 = R$		$28.5 = R$		$26.5 = R$

$N = 5 + 5 + 5 = 15$

The final value after correction is $H = 9.8$. This study has 3 groups minus 1 or 2 degrees of freedom (df). Referring to the chi square table, the value needed for 2 df at the .05 level of significance is 5.99. Since the H value of 9.8 is greater than 5.99, we conclude that there is a difference between the groups. We then logically surmise that this difference is in the alcohol-drinking driver because of the higher number of errors he has incurred as compared to the other two drivers.

Problems

7.1 A small corporation has two manufacturing departments. Department A has a substantially superior safety record compared to Department B. Management believes that the difference is attributable to a more positive attitude toward the corporation. To prove this, management has purchased an Attitude Inventory from a management consulting firm. The inventory's scores vary from 0% (very poor) to 100% (excellent). Almost all production workers in each department were tested. The results appear in Table 7.7. Compare the test scores and state a conclusion.

7.2 Continuing with problem 7.1, management gives a 10-minute "pep talk" to Department B in the cafeteria. The next day, the same inventory is readministered to Department B. The results appear in Table 7.8. Compare these test results with the prior test results. What is your conclusion?

Table 7.7. Inventory Results

Department A	Department B		Department A	Department B
97	89		87	71
82	83		88	68
84	78		86	67
91	79		88	72
95	65		81	70
95	69		80	75
90	76		96	77
88	73			66

Table 7.8. Inventory Retests

Department B
60
90
84
74
64
78
86
82
67
81
77
69
59
63
78
68

Table 7.9. Height in Centimeters

Inner City Workers	Midwest Workers	Inner City Workers	Midwest Workers
160	168	155	159
163	172	160	174
156	170	164	171
161	167	155	175
162	163	153	169

7.3 Emma is a corporation ergonomist who has been asked to design a large number of workstations for a new plant that her corporation is setting up in Inner City. Many of the workers in the new plant will be immigrants. Emma has already designed workstations for company plants in the Midwest. She is concerned that the present Midwest workstation dimensions will not apply to the Inner City requirements. Emma wishes to make comparisons between the two different workforces by obtaining stature information samples from records of company physicals. The results appear in Table 7.9. Is there a difference between the two groups? Which way?

7.4 Mike is a mechanical engineer who is designing a guard for a table saw. He has created three possible designs and wishes to find out which of them will withstand the greatest abuse. He subjects groups of five of each of the three prototype guards to abuse tests and counts the number of trials needed before the guard breaks. The data appear in Table 7.10. Statistically determine the difference between guard testing results. Which is the most reliable guard?

Table 7.10. Trials Needed to Break Guard (in thousands)

Guard 1	Guard 2	Guard 3
1.55	3.00	1.70
.90	2.80	1.05
1.30	2.95	1.80
2.00	3.20	.75
.50	2.70	1.20

Chapter 8

Random Samples, Large and Small

8.1 Introduction. In prior chapters, and in the following chapters of this book, it is assumed that observations are made about groups that are chosen ad hoc and expeditiously and not through any difficult time-consuming formalized procedures such as random sampling. Safety professionals are likely to select groups by expeditious, nonrandomized procedures because of time restraints, expense, and other practical considerations. This ad hoc choosing of groups leads to the use of nonparametric statistics, which are those usually found in this book. If, however, reasonably large groups are chosen through random sampling, then these groups are likely to conform to the patterns of normal distribution. This in turn will allow the use of parametric statistics and increase the likelihood of drawing important statistical conclusions. This chapter will describe two major parametric techniques: confidence intervals and comparison of means.

8.2 Confidence Intervals. If a reasonably large random sample is tested, a *mean* (average) may be calculated in order to derive a number that represents the entire group of data. Because of sampling variations, the mean derived from the testing will not be totally reliable; thus, a *confidence interval* may be provided with the mean. This procedure indicates the probability that the mean, which is stated within a bracketed set of numbers, is a correct representation of the universe that has been sampled. The testing procedure used upon the group should produce a number located on a scale of equally increasing intervals. An example follows.

8.2.1 Case Study. In the past several years, a large number of new employees have been given a questionnaire after receiving their initial safety training. One question on the questionnaire asks the new employee to indicate how valuable the training was by giving it a score from 1 to 10. From this large number of questionnaires, 50 of the responses about the value of safety training are randomly chosen. The results are shown in Table 8.1. The mean score ($\overline{X}$) is 306 divided by 50 = 6.12.

To calculate the confidence interval for this mean, each score (labeled X) is listed. Then each score (X) is squared (X^2) as in Table 8.2, and all of the squared

Table 8.1.   Responses about Value of Safety Training

4	5	1	8	5	6	10	9
6	3	4	6	10	9	8	7
7	3	2	4	5	7	6	8
9	6	8	5	2	6	5	9
10	8	6	5	4	7	6	8
2	9	1	3	10	9	7	6
7	5						

Table 8.2.   Responses about Value of Safety Training

X	X^2	X	X^2	X	X^2	X	X^2	X	X^2
4	16	6	36	8	64	4	16	6	36
6	36	8	64	6	36	10	100	5	25
7	49	9	81	4	16	6	36	6	36
9	81	5	25	5	25	9	81	7	49
10	100	1	1	5	25	7	49	9	81
2	4	4	16	3	9	6	36	7	49
7	49	2	4	5	25	7	49	8	64
5	25	8	64	10	100	9	81	9	81
3	9	6	36	5	25	10	100	8	64
3	9	1	1	2	4	8	64	6	36
									2,168

quantities are added together. Finally, a formula used to compute a statistic known as the *standard error* is applied. The calculation proceeds as follows:

X = individual score N = number of scores = 50

X^2 = score squared ΣX^2 = 2,168

$\bar{X}$ = mean score = 6.12

$$\text{Standard error (SE)} = \frac{\sqrt{\dfrac{\Sigma X^2}{N} - \bar{X}^2}}{\sqrt{N-1}}$$

$$\text{SE} = \frac{\sqrt{\dfrac{2,168}{50} - 6.12^2}}{\sqrt{50-1}}$$

$$\text{SE} = \frac{\sqrt{43.36 - 37.45}}{\sqrt{49}}$$

$$\text{SE} = \frac{2.43}{7} = .347$$

$$df = N - 1 = 50 - 1 = 49$$

t value of 49 at .05 level = 2.008

Confidence interval = $\bar{X} \pm 2.008(\text{SE})$

Confidence interval = $6.12 \pm 2.008(.347)$

Confidence interval = $6.12 \pm .70$

Confidence interval = 5.42 to 6.82

The standard error is determined to be .347. In order to derive the confidence interval, the standard error must be multiplied by a value from Table 2 in Appendix D and added to and also subtracted from the mean. To determine this Table 2 value, the degrees of freedom (df) must be calculated: $df = N - 1 = 50 - 1 = 49$. At the value closest to 49 df (50 df) in Table 2, the table value (t) is 2.008 at the .05 (5%) level of significance and 2.678 at the .01 (1%) level of significance. If working at the .05 level, the confidence interval is $\bar{X}$ (612) $\pm$

t (2.008) × SE(.347). The final statement made by the researcher is that there is a 95% probability that the mean score lies between 5.42 and 6.82.

8.2.2 Discussion. This procedure is applicable to smaller samples as well as large, randomly chosen samples. It must be emphasized that the t value from Table 2 must be correctly chosen by calculating the *df* of the study at hand. Note that the larger the random sample collected, the narrower the confidence interval should be.

8.3 Differences between Means. Another procedure enables us to determine the statistical difference between large or small randomly drawn samples of equal or unequal size. It is necessary that both of the samples are tested on an equal interval scale. An example of a comparison of two small samples of unequal size follows.

8.3.1 Case Study. An ergonomist has given a grip strength test to a randomly chosen group of smaller-than-average men and a randomly chosen group of larger-than-average women. Contrary to popular expectations, the women had a higher average grip strength than the men. The ergonomist decides to determine whether this is a usual occurrence or just a chance occurrence. Using the data in Tables 8.3 and 8.4, the computation for comparing the two groups proceeds as follows:

$$\text{Men } (\overline{X}_1) = 61.88 \qquad \text{Note: } X = \text{individual score}$$

$$\text{Women } (\overline{X}_2) = 89.5 \qquad \overline{X} = \text{mean (average)}$$

$$\text{Men: } \Sigma\, X_1^2 = 34{,}597 \qquad N = \text{number of sample scores}$$

$$\text{Women: } \Sigma\, X_2^2 = 64{,}124 \qquad \Sigma\, X^2 = \text{sum of scores squared}$$

$$\text{Men: } N_1 = 9$$

$$\text{Women: } N_2 = 8$$

Step 1:	Standard deviation (SD) $= \sqrt{\dfrac{\Sigma X^2}{N} - \overline{X}^2}$

$$\text{Men: } SD_1 = \sqrt{\frac{34{,}597}{9} - 61.88^2} \qquad \text{Women: } SD_2 = \sqrt{\frac{64{,}124}{8} - 89.5^2}$$

$$SD_1 = \sqrt{3{,}844.11 - 3{,}829.13} \qquad SD_2 = \sqrt{8{,}015.5 - 8{,}010.25}$$

$$SD_1 = 3.87 \qquad SD_2 = 2.29$$

Step 2:   Standard error (SE) = $\sqrt{\dfrac{N_1(\text{SD}_1^2) + N_2(\text{SD}_2^2)}{N_1 + N_2 - 2}\left(\dfrac{1}{N_1} + \dfrac{1}{N_2}\right)}$

$$\text{SE} = \sqrt{\dfrac{9(3.87^2) + 8(2.29^2)}{9 + 8 - 2}\left(\dfrac{1}{9} + \dfrac{1}{8}\right)}$$

$$\text{SE} = \sqrt{\dfrac{134.79 + 41.95}{15}(.111 + .125)}$$

$$\text{SE} = \sqrt{11.78(.236)}$$

$$\text{SE} = 1.67$$

Step 3: t value $= \dfrac{|\bar{X}_1 - \bar{X}_2|}{\text{SE}}$

$$t \text{ value} = \dfrac{|61.88 - 89.5|}{1.67}$$

$$t \text{ value} = \dfrac{27.62}{1.67} = 16.54$$

Degrees of freedom $= df = (N_1 - 1) + (N_2 - 1) = (9 - 1) + (8 - 1) = 15df$

Table 8.3. Grip Strengths in Pounds of Force

Men	Women
61	92
68	91
59	88
55	90
62	86
60	93
65	87
66	89
61	$\overline{716}$
$\overline{557}$	

Average $(\bar{X})$ for men $= 557 \div 9 = 61.88$
Average $(\bar{X})$ for women $= 716 \div 8 = 89.5$

Table 8.4.   Grip Strengths in Pounds of Force

X	X^2	X	X^2
61	3,721	92	8,464
68	4,624	91	8,281
59	3,481	88	7,744
55	3,025	90	8,100
62	3,844	86	7,396
60	3,600	93	8,649
65	4,225	87	7,569
66	4,356	89	7,921
61	3,721		64,124
	34,597		

The comparison between the men's group and the women's group requires three steps. In Step 1, a statistic known as the *standard deviation (SD)* is determined for each group. The SD is derived by using the sum of the squares of the individual strength scores, the number of strength scores, and the mean. Step 2 is a determination of a combined standard error (SE) for both groups using both standard deviations and the number of scores. In Step 3 a *t* value is determined by calculating the absolute difference of both means and dividing by the standard error. The final *t* value is 16.54. To refer this value to Table 2 in Appendix D, the degrees of freedom (*df*) must be found: $df = (N - 1) + (N-1) = (9 - 1) + (8 - 1) = 15$ *df*. To use this statistic, the ergonomist assumes that there is no difference between the two groups. The *t* value for 15 *df* in Table 2 is 2.131 at the .05 (5%) level of significance and 2.947 at the .01 (1%) level of significance. Since the determined *t* value of 16.54 exceeds the table values of 2.131 and 2.947, the assumption that there is no difference between the groups is not correct. The ergonomist therefore concludes that there is a more than 95% probability and even a more than 99% probability that samples similar to this would be generated again and further concludes that large women have a stronger gripping strength than small men.

8.3.2 Discussion. It must be remembered that the testing procedure for both groups must be on a scale of equal intervals and that the samples of both groups must be randomly chosen. Note also that the larger the sample size of the groups and the more homogeneous the scores and the more closely they cluster around the means, the greater the likelihood that a statistical difference will occur between the samples.

Problems

8.1 A human factors engineer is interested in the reaction time of a randomly chosen group of truck drivers. Each driver was asked to press a lever when a green light appeared on a board. Reaction time testing equipment measured the time that expired between the appearance of the green light and the activation of the lever. The results appear in Table 8.5.

  a. Compute a mean reaction time for the group.
  b. Compute a confidence interval at the 5% level of significance.
  c. Compute a confidence interval at the 1% level of significance.

8.2 The human factors engineer adds another variable to the above study. A green or red light will now be displayed to another sample of drivers. The drivers are told to press the lever only when the red light appears. Table 8.6 presents the results.

  a. Compute a mean reaction time for the group.
  b. Compute a confidence interval at the 5% level of significance.
  c. Compute a confidence interval at the 1% level of significance.
  d. Make a statistical comparison between this group and the group in problem 8.1.

Table 8.5.   Simple Reaction Time (in Milliseconds)

209	195	210	201
200	198	199	202
207	191	200	209
219	191	210	196

Table 8.6.   Choice Reaction Time (in Milliseconds)

310	315	315
303	299	299
311	313	308
314	317	301

8.3 Jones is the safety officer of a large shopping mall. The exterior walkways of the older section of the mall are composed of highly glazed ceramic tile. The exterior walkways of the new section are composed of quarry tile with a rough finish. Jones is interested in the relative slip resistance to pedestrians of both types of tile. He takes surface friction tests with a standard slip meter for randomly chosen areas of both glazed tile and quarry tile. The Static Coefficient of Friction (SCOF) readings that the meter might produce vary theoretically from 0 (extremely slippery) to 1.0 (extremely slip resistant). The readings appear in Table 8.7.

a. Calculate the means and confidence intervals of both samples.
b. Statistically compare the two samples.

Table 8.7. Static Coefficient of Friction Readings

New Walkways	Old Walkways
.46	.67
.50	.68
.48	.71
.43	.65
.47	.72
.45	.68
.51	.70
.50	.66
.46	.73
.42	.70

Chapter *9*

Numeric Indicators

9.1 Background. In some situations, a researcher may decide to use a single number to represent a category or variable. This number may be a percentage, a ratio, an index, or a rate. This number may then be compared with the *numeric indicators* of other categories or variables, or it may be compared against itself over time. Numeric indicators, such as the highly publicized Consumer Price Index (CPI), are frequently used in economics. The science of psychology uses the well-known IQ (intelligence quotient) as a numeric indicator. Numeric indicators are often used in safety and health to describe accident and illness occurrences and may be used in other safety and health areas as well. The procedure for deriving safety and health indicators will be addressed in a later section.

9.2 General Method. A numeric indicator is usually a decimal number that is the result of an arithmetic division between two other numbers. The two numbers are stated as a fraction. If the numerator (upper number) of the fraction is larger than the denominator (lower number), the resulting decimal number is greater than 1. If the numerator is smaller than the denominator, the resulting decimal is less than 1. To obtain a numeric indicator that is a whole number and is of manageable size, the decimal number is often, but not necessarily, multiplied by 100 or some other larger number.

9.2.1 Examples. A fraction has a numerator of 12 and a denominator of 10: $12 \div 10 = 1.2$. If desired, the resulting indicator of 1.2 may be multiplied by 100, producing the more manageable number of 120. Likewise, if the numerator is 10 and the denominator is 12, the indicator is .83, which becomes 83 when multiplied by 100. Finally, note that an identical numerator and denominator produce an indicator of 100.

9.3 Types of Numeric Indicators. There are many varieties of numeric indicators, particularly in the fields of economics and business. We will cover only some methodologies that are used or may be used in safety and health.

9.3.1 Percentages. *Percentages* have already been discussed in section 2.4 as convenient representations for categories and variables. Percentages are also well suited for facilitating comparisons between groups of unequal size. Table 9.1 illustrates this point. The safety manager of Smith Trucking has discovered that during the last year, 22 of her company's truck drivers have received at least one traffic violation. She has also discovered from her friend, who is the safety manager of Acme Trucking, that 24 Acme drivers received moving violations in the same period of time. Upon hearing this, the safety manager is relieved that Smith's incidence of traffic violations approximately matches that of another company and thus can't be very bad. Actually, she has been lulled into complacency. It has not occurred to her that Acme has almost twice as many drivers as Smith. If the safety manager had used percentages as found in Table 9.1, she would have discovered that her company's violation record is possibly worse than that of Acme.

9.3.2 Ratios. *Ratios* are a statement about the relationship between the parts of a whole. Another way to compare Smith and Acme is to compute the relationship of nonviolating drivers to violating drivers as in Table 9.2. This

Table 9.1. Percentages of Drivers with Moving Violations

	Acme Trucking	Smith Trucking
Total number of drivers	183	98
Drivers with moving violations	24	22
Percentage of drivers with moving violations	13%	22%

Table 9.2. Ratios of Nonviolating Drivers to Violating Drivers

Drivers	Acme	Smith
Nonviolating drivers	159	76
Violating drivers	24	22

Acme ratio of nonviolators to violators: $159 \div 24 = 6.6:1$ (rounded to 7:1)

Smith ratio of nonviolators to violators: $76 \div 22 = 3.5:1$ (rounded to 4:1)

again indicates that Smith may have a worse record than Acme because Acme has more nonviolating drivers per violating drivers than Smith.

9.3.2.1 Comparisons between Groups. A statistical test should be performed between two groups in order to compare their respective percentages or respective ratios. To perform this test, both the size of each group and the percentage or ratio of at least one variable of each group must be known. The following calculation is based upon the data for the violating and nonviolating drivers of Acme Trucking and Smith Trucking. The technique is the chi square technique that we used in Chapters 3, 4, and 5. Note that the sum of the ratio numbers (in this case 6.6:1 = 7.6 and 3.5:1 = 4.5) divided into the group size produces the size of the smaller variable.

Acme: Total drivers = 183; violating drivers = 13%.
Smith: Total drivers = 98; violating drivers = 22%.
Acme: 183(13%) = 24 violating drivers; 183 − 24 = 159 nonviolating drivers.
Smith: 98(22%) = 22 violating drivers; 98 − 22 = 76 nonviolating drivers.

These data can also be stated as follows:

$$\text{Acme: Nonviolators to violators} = 6.6{:}1 = \frac{183}{6.6 + 1} = 24 \text{ violating drivers}$$

$$\text{Smith: Nonviolators to violators} = 3.5{:}1 = \frac{98}{3.5 + 1} = 22 \text{ violating drivers}$$

Applying the chi square (χ^2) technique to these data, we have the following:

	Nonviolators	Violators	
Acme	159 (*A*)	24 (*B*)	*A* + *B* = 183
Smith	76 (*C*)	22 (*D*)	*C* + *D* = 98
			A + *C* = 235
			B + *D* = 46
			AD = 3,498
			BC = 1,824
			N = 281

$$\chi^2 = \frac{N\left(|AD - BC| - \frac{N}{2}\right)^2}{(A + B)(C + D)(A + C)(B + D)}$$

$$\chi^2 = \frac{281\left(|3{,}498 - 1{,}824| - \frac{281}{2}\right)^2}{(183)(98)(235)(46)}$$

$$\chi^2 = 3.41$$

This value of 3.41 does not exceed the χ^2 value of 3.84 at 1 df at the .05 level. Thus, although the Smith percentages and ratios do not appear favorable compared to Acme, there is no difference between the two companies.

9.3.3 Index Numbers. *Index numbers* are used to represent a category or variable as it changes over time. Index numbers are calculated as described in section 9.2. An example of their use follows.

Mr. Moreau is the risk manager of an industrial corporation. One of his most important responsibilities is purchasing insurance at reasonable prices. He is particularly concerned about the size of the annual worker's compensation insurance premiums. He has been accumulating data for the last 11 years as shown in Table 9.3.

Table 9.3. Annual Worker's Compensation Premiums

Year	Premium (Thousands of Dollars)	Index (Base 1986) (Multiplied by 100)
1986	19.0	100
1987	18.5	97
1988	18.1	95
1989	19.2	101
1990	22.8	120
1991	24.2	127
1992	24.6	129
1993	25.0	132
1994	25.1	132
1995	25.4	134
1996	25.2	133

In 1986 Mr. Moreau decided to track his worker's compensation insurance expenditures over time by calculating index numbers for each year's premium payments. To do this, he had to determine a base year for the computations. A base time interval for index computations is often found at the beginning of the time series. Other intervals may be used, but there should be a good reason for their choice such as typicality, recency, all-time low or all-time high, and the like. Since Mr. Moreau was starting out with this study, he had no choice but to select 1986 (the beginning year) as his base year. As the years pass, he divides each year's premium payment by the base year (1986) premium payment and then multiplies by 100 to arrive at a more manageable number. Table 9.3 contains index numbers for each year based upon 1986. The table indicates that expenditures were level and low in the 1980s and have been rising ever since. From this table, Mr. Moreau should be able to determine how effective his worker's compensation insurance purchases are. Note also that Chapter 13's time series techniques, which are related to graphing and trend analysis, may be applied concurrently to this indexing procedure.

9.3.3.1 Change of Base Year. Mr. Moreau chose 1986 as his base year. As the years pass, if he wished, he could change to another base year. He might, for example, change the base from 1986 to 1990, when the insurance premiums underwent an abrupt long-term general increase in size due to legislation at the state level. Changing the base simply requires a recalculation of each year according to the new base. This recalculation appears in Table 9.4.

Table 9.4.   Annual Worker's Compensation Premiums

Year	Premium (Thousands of Dollars)	Index (Base 1990) (Multiplied by 100)
1986	19.0	83
1987	18.5	81
1988	18.1	79
1989	19.2	84
1990	22.8	100
1991	24.2	106
1992	24.6	108
1993	25.0	110
1994	25.1	110
1995	25.4	111
1996	25.2	111

9.3.3.2 Aggregate Index Numbers. The index numbers described in the prior sections were the result of calculations performed upon a single category or variable. By using simple addition, index numbers may be created by combining more than one variable. Adding together several variables may provide a more universal, meaningful index number. It must be remembered, however, that the variables being combined must be logically related to each other. An example related to aggregating numbers appears in Table 9.5.

Mr. West is the safety manager of a large chain of supermarkets. The company has a large number of retail clerks checking out groceries. The company also has a substantial number of clerks in various regional offices working on visual display terminals (VDTs). Mr. West is aware of the problem of carpal tunnel syndrome (CTS) as related to these two occupations and is convinced that many hours expended at these jobs will lead to a large number of compensation claims. He started an aggregate indexing system six years ago calculated upon hours expended by both types of workers using the first year (1992) as a base. An aggregate index for a year equals retail clerk hours plus VDT operator hours divided by the base year's retail clerk plus VDT operator hours multiplied by 100. This aggregate index, which takes two variables into consideration, indicates to Mr. West, as he computes his yearly indexes, that there is a slow but steady rise in hours expended in occupations that are often causal to CTS.

9.3.3.3 Weighted Aggregates. The aggregate method described in the previous section is an effective procedure in that it attempts to give consideration to all pertinent variables related to the index number at hand. This procedure, however, assigns equal impact and importance to all variables concerned. Mr. West can remedy this situation in the following manner. Searching through company medical records, he discovers that the VDT operators make 2.5 more CTS claims per worker than the retail clerks do. He decides to give extra weight

Table 9.5.   Production Hours of Retail Clerks and VDT Operators (in 100,000s)

Year	Retail Clerk	VDT Operator	Total Hours	Aggregate Index (Base 1992) (Multiplied by 100)
1992	4.7	.9	5.6	100
1993	4.8	1.1	5.9	105
1994	4.5	1.2	5.7	101
1995	4.7	1.3	6.0	107
1996	4.5	1.5	6.0	107
1997	4.4	1.5	5.9	105

and importance to the hours put in by VDT operators by multiplying their hours by 2.5. In Table 9.6 each index number is derived by multiplying retail clerk hours by 1 and adding this product to the product of VDT operator hours multiplied by 2.5; then this sum is divided by the base year, and the result is multiplied by 100. From Table 9.6 it is even more apparent that hours put into CTS-prone activities are on the increase.

9.3.4 Rates. *Rates* are ratios that deal with the change of a category or variable over a particular unit of time. As an example, if an automobile travels 120 miles over a period of 2 hours, then the auto's rate of movement across space is 120 miles divided by 2 hours or 60 miles per hour. Similarly if a corporation's workforce increased by 1,200 over a period of 12 months, then the growth rate of personnel is 1,200 workers divided by 12 months or 100 workers per month. Finally, if the total number of a company's unabated internally written safety violations decreases by 220 over 4 years, then the rate of decrease of safety violations is 220 violations divided by 4 years or 55 violations per year. Note that such rates are likely to be average rates and that the change was probably not uniform over the period of time involved. In the example of the automobile, it is very possible that velocity varied and that the auto traveled, say, 70 mph in the first hour and 50 mph in the second hour but still produced a rate of 60 mph.

It is also important to note that in these examples the rate was a result of division within a fraction and that the numerator was always larger than the denominator. The reverse may sometimes occur. Changing the moving object in the first example to a tortoise, we find that it travels 2 miles over a period of 120 hours. In this case, the tortoise's rate is 2 miles divided by 120 hours or .017 miles per hour. This number may be made whole and more manageable by

Table 9.6. Production Hours of Retail Clerks and VDT Operators (in 1,000s)

Year	Retail Clerks	Weight	VDT Operators	Weight	Total	Index Number (Base 1992) (Multiplied by 100)
1992	4.7	1	.9	2.5	6.95	100
1993	4.8	1	1.1	2.5	7.55	109
1994	4.5	1	1.2	2.5	7.50	108
1995	4.7	1	1.3	2.5	7.95	114
1996	4.5	1	1.5	2.5	8.25	119
1997	4.4	1	1.5	2.5	8.15	117

multiplying by a large number such as 1,000. The tortoise's progress may be more conveniently calculated as 2 miles times 1,000 divided by 120 hours, producing a rate of 17 miles per 1,000 hours.

9.3.4.1 Safety and Health Rates. The safety and health profession is particularly interested in monitoring the effectiveness of its own programs. Accident and/or illness rates calculated according to the previously described techniques provide two very helpful contributions: (1) These rates, if regularly calculated, may allow comparisons over time. Thus, today's rates, compared to prior rates, are a good indicator of the success of an organization's safety efforts. (2) These rates allow for comparisons between groups and organizations of unequal size and unequal characteristics to determine which safety efforts are more or less successful and which groups may have more inherently hazardous operations. These rates fall into two major categories: (1) *incidence* (also called *frequency*) *rates,* which consider the number of injuries and/or illnesses occurring within a particular period of time (usually a year); and (2) *severity rates,* which consider the negative consequences of these injuries or illnesses, usually by counting the number of days the injured or sick workers were away from the job during a particular period (usually a year).

9.3.4.2 Safety and Health Rate Calculations. Incidence rates are calculated by dividing the number of injuries and/or illnesses within a period of time by the hours expended by the group or organization within the same period of time. Thus, if a small company experienced 21 injuries and illnesses during a year and its employees worked 90,000 hours during the year, its incidence rate is .00023, calculated by dividing 21 injuries by 90,000 hours.

Severity rates are calculated by dividing the number of days lost from the job because of these injuries and/or illnesses by the hours expended by the group or organization within the same period of time. Thus, if the same company experienced 68 lost days during the same year, then the severity rate is .00076, calculated by dividing 68 days by 90,000 hours.

Usually, the numerator in these calculations is substantially smaller than the denominator, so a larger number is multiplied against the numerator to create a more manageable result. OSHA (Occupational Safety and Health Administration) and the BLS (Bureau of Labor Statistics) suggest/require multiplying by 200,000. ANSI (American National Standard Institute) at one time suggested multiplying by 1,000,000. Figure 9.1 demonstrates various calculations related to these procedures. It should be remembered that an organization may choose these procedures or any variation upon these calculations it deems fit for its own internal needs. For any calculations that might be required by outside agencies such as OSHA or the BLS, the current National Safety Council's "Accident Prevention Manual for Industrial Operations" should provide adequate additional information.

Incidence (Frequency) Rate Using 200,000	**Severity Rate Using 200,000**
$$\frac{\text{Number of injuries \& illnesses} \times 200,000}{\text{Total hours worked}} = R$$	$$\frac{\text{Number of days lost} \times 200,000}{\text{Total hours worked}} = R$$
$$\frac{21 \times 200,000}{90,000} = 46.7$$	$$\frac{68 \times 200,000}{90,000} = 151.1$$
Incidence (Frequency) Rate Using 1,000,000	**Severity Rate Using 1,000,000**
$$\frac{\text{Number of injuries \& illnesses} \times 1,000,000}{\text{Total hours worked}} = R$$	$$\frac{\text{Number of days lost} \times 1,000,000}{\text{Total hours worked}} = R$$
$$\frac{21 \times 1,000,000}{90,000} = 233.3$$	$$\frac{68 \times 1,000,000}{90,000} = 755.6$$

Total hours worked in one year:	90,000
Number of injuries and illnesses in same year:	21
Resulting days lost from the job in same year:	68

Figure 9.1.   Safety and Health Rate Calculations

9.3.4.3 Comparing Frequency Rates. A statistical comparison can be made between two frequency rates. This comparison can be made as part of a before with control design (see Chapter 3) or as part of a before and after design (see Chapter 4). To make this comparison, the basic fractions upon which the rates are constructed must be made available by either inquiry or record search.

9.3.4.3.1 Example. The safety officers of two similar manufacturing firms wish to compare their latest frequency rates. In revealing their rates to each other, they must also reveal the calculations needed to derive these rates. Note that in this example, the safety officers use different multiplying numbers within their numerators, but this problem can be bypassed by working with the original fraction. The calculation follows:

Corporation 1 frequency rate:

$$\frac{22(1,000,000)}{92,500} = 237.84$$

Corporation 2 frequency rate:

$$\frac{25(200,000)}{95,750} = 52.22$$

Basic fraction: $\dfrac{A_1}{H_1} = \dfrac{22}{92,500}$

Basic fraction: $\dfrac{A_2}{H_2} = \dfrac{25}{95,750}$

Comparison formula:

$$\dfrac{\left| \dfrac{A_1}{H_1} - \dfrac{A_2}{H_2} \right|}{\sqrt{\left(\dfrac{A_1 + A_2}{H_1 + H_2} \right) \left(1 - \dfrac{A_1 + A_2}{H_1 + H_2} \right) \left(\dfrac{1}{H_1} + \dfrac{1}{H_2} \right)}}$$

$$\dfrac{\left| \dfrac{22}{92,500} - \dfrac{25}{95,750} \right|}{\sqrt{\left(\dfrac{22 + 25}{92,500 + 95,750} \right) \left(1 - \dfrac{22 + 25}{92,500 + 95,750} \right) \left(\dfrac{1}{92,500} + \dfrac{1}{95,750} \right)}}$$

$$\dfrac{.00002326}{\sqrt{(.0002497)(.99975)(.00002125)}}$$

$$\dfrac{.00002326}{.00007283} = .32$$

The result of the calculation is .32. In this procedure, statistical significance is 1.96 at the .05 level of probability and 2.58 at the .01 level of probability. The resulting quantity of .32 does not reach these values; thus, the safety officers conclude that there is no difference between these two rates, and the safety performance of both companies is judged to be about the same.

9.3.4.3.2 Another example. A safety officer wishes to compare his current frequency rate with last year's rate. The calculation follows:

This year's frequency rate: Last year's frequency rate:

$$\dfrac{31(1,000,000)}{98,600} = 314.40 \qquad\qquad \dfrac{55(1,000,000)}{94,350} = 582.94$$

$$\dfrac{\left| \dfrac{31}{98,600} - \dfrac{55}{94,350} \right|}{\sqrt{\left(\dfrac{31 + 55}{98,600 + 94,350} \right) \left(1 - \dfrac{31 + 55}{98,600 + 94,350} \right) \left(\dfrac{1}{98,600} + \dfrac{1}{94,350} \right)}}$$

$$\dfrac{.0002685}{\sqrt{(.000446)(.999554)(.00002074)}}$$

$$\dfrac{.0002685}{.00009616} = 2.79$$

The result of the calculation is 2.79. This exceeds the .05 level of probability of 1.96 and even the .01 level of 2.58. Thus, the safety officer concludes that the two rates are different and that the company safety performance has improved over the past year.

9.3.4.4 Combined Rates. There is no complete agreement within the profession as to whether incidence rates or severity rates are more important. A high current severity rate is always a concern because of the human suffering involved as well as the effects upon company profits. A low current incidence rate, though impressive in appearance, may still be of concern if the few incidents that occurred caused many days to be lost to the organization. Likewise a high incidence rate may potentially cause a high severity rate in the future even though the current severity rate is low. The most desirable situation is a low incidence rate as well as a low severity rate.

Some organizations try to develop a single number that reflects both rates. This is done by multiplying the current incidence rate by the current severity rate. If both current rates are low, then their product will also be low, indicating a highly desirable situation. This procedure is referred to as the *incidence/severity ratio* or the *frequency/severity ratio*. Using the calculations in Figure 9.1, this ratio would be either $46.7 \times 151.1 = 7,056$ or $233.3 \times 755.6 = 176,282$. To make these numbers more manageable, the first form of the ratio might be divided by 1,000, producing 7.1, and the second form might be divided by 10,000, producing 17.6. If this procedure of making more manageable numbers is used, then it must be faithfully repeated from time period to time period.

Problems

9.1 Mr. James, who is in charge of safety training for his corporation, has been using the services of three outside private training firms to provide a training course for company forklift drivers. After training is completed, the drivers are all given the same test, which was designed by Mr. James. The test is difficult and not all of the drivers pass the first time. The number of drivers trained by each of the training firms varies. Mr. James wishes to test the effectiveness of these firms and collects the information in Table 9.7.

a. Construct a table and compute percentages to determine which firm has the best training record.

Table 9.7. Company Drivers Taking Forklift Training

	Firm 1	Firm 2	Firm 3
Total number trained	124	43	58
Number who failed	6	7	8

b. Construct a table and compute ratios between passing and failing drivers to determine which firm has the best training record.

9.2 The XYZ Insurance Corporation has been insuring the doctors and nurses of a prominent hospital for professional liability. The hospital has a total of 158 full-time doctors and 432 full-time nurses. Since the inception of the policy, 38 doctors and 73 nurses have been sued at least once for professional errors or omissions.

a. Construct a table and compute percentages to determine which profession has the better record.
b. Construct a table and compute ratios to determine which profession has the better record.

9.3 The National Safety Council has provided some data related to pedalcycle accidents, which are shown in Table 9.8.

a. Using 1940 as the base, construct index numbers for all years. What can you conclude?
b. Changing the base to 1955, construct index numbers for all years. Was this an advisable change?

9.4 A large nationwide construction company has large annual expenditures on personal protective equipment, especially hard hats and safety shoes. The purchasing agent for the corporation is interested in controlling

Table 9.8. Pedalcycle Death Rates (per 100,000 Pedalcyclists)

Year	Rate
1940	9.59
1945	5.55
1950	3.18
1955	1.78
1960	1.63
1965	1.75
1970	1.38
1975	1.05
1980	1.46

Table 9.9. Expenditures per 100 Construction Workers (in Thousands of Dollars)

Year	Hard Hats	Safety Shoes
1988	20.1	29.0
1989	20.1	29.0
1990	20.0	29.1
1991	19.8	29.7
1992	19.9	29.8
1993	19.5	30.0
1994	19.3	30.9
1995	19.0	31.0
1996	18.3	31.2

expenditures for these items and has been tracking expenses over the years. These data appear in Table 9.9.

a. Using 1988 as a base year, construct a table containing aggregate index numbers for each year.
b. What is the general trend in personal protective equipment expenditures? How does this trend relate to individual expenditures for hard hats as well as safety shoes?

9.5 A company has a large fleet of sales representatives' autos and an equally large fleet of delivery trucks. The company has been keeping accident records for more than eight years. Table 9.10 lists the number of collisions that involved $500 damage or more for autos and trucks in each year.

a. Using the first year as a base, construct a table containing aggregate index numbers for each year.
b. What is the general trend of collision incidence over the years?

9.6 Refer to Table 9.10. Research into company records indicates that a typical truck collision causes twice as much property damage as a typical auto collision.

a. Construct a table containing weighted aggregate index numbers for each year.

Table 9.10. Number of Collisions
Involving Damage of $500 or More

Year	Autos	Trucks
1990	25	15
1991	22	16
1992	24	17
1993	21	18
1994	20	18
1995	19	20
1996	19	22
1997	16	22

b. Does this change the collision trends from those determined in the prior problem?

9.7 For the current year, a company has determined the following information from its injury and illness and payroll records:

- Number of on-the-job injuries and illnesses: 66
- Days away from the job because of these injuries and illnesses: 305
- Total hours worked by all employees during the year: 953,889

a. Calculate incidence rates and severity rates from the above information using the standard multiplier requested by OSHA and the BLS.
b. Do the same for the above information using the standard multiplier of ANSI.
c. Devise a standard multiplier of your own that you think would best suit this company's needs.
d. Calculate combined ratios from the above rates. Simplify these rates with a standard divisor that you believe would best suit the company's needs.

9.8 Compare the 1996 frequency rates of Ajax Corporation and Achilles Corporation. Ajax expended 105,150 work hours and experienced 56 reportable injuries and illnesses. Its frequency rate is 532.58. This is based upon a multiplier of 1,000,000. Achilles expended 201,325 work hours and experienced 65 reportable injuries. Its frequency rate is 64.57. This is based upon a multiplier of 200,000.

9.9 Last year (1995) Achilles worked 203,580 hours and experienced 85 reportable injuries. Compare 1995 with 1996.

Chapter 10

Other Statistical Tests

10.1 Background. Two additional nonparametric tests based upon the chi square (χ^2) may be of use to the researching safety and health professional. These techniques are unique yet similar to those used in Chapters 3, 4, 5, and 8.

10.2 Testing against a Distribution. This test is suited for making comparisons within a group, sample, or population and may also allow comparisons with other groups. It may be used generally in descriptive statistics and with the before only design described in Chapter 2. The important concept that pervades this technique as well as other χ^2 techniques is the comparison between what is observed about a group and what could be theoretically expected of a group.

10.2.1 Example: Sampling Based upon a 50:50 Distribution. Jack is the safety officer of a corporation whose operations include a great deal of milling and grinding. The work environment contains a large number of machines producing flying particles, and it is imperative that workers wear safety glasses. Many workers refrain from wearing the glasses, however, and complain that the present glasses are unfashionable and unattractive. Jack is willing to purchase more expensive and what he believes are more attractive safety glasses provided that the workers approve of them. He decides to poll all 190 members of a typical department in the company to determine whether they approve of a certain new design. He hopes that their response will help him make a conclusion about the rest of the company's workers. His findings are shown in Table 10.1.

Jack originally stated that he would be willing to purchase the new glasses if more than 50% of all workers approved. Currently, 60% (114 workers) of one department approve. He is concerned, however, that the results of this sample

Table 10.1. Workers' Response to
New Safety Glasses Design

Approve	Disapprove	Total
114	76	190

Table 10.2. Observed and Expected Responses to the New Design

	Approve	Disapprove	Total
Observed (O)	114	76	190
Expected (E)	95	95	190

Table 10.3. Chi Square Determination

	Approve	Disapprove	Total
O	114	76	190
E (50:50)	95	95	190
$O - E$	+19	−19	
$(O - E)^2$	361	361	
$\dfrac{(O-E)^2}{E}$	3.8	3.8	7.6

$$\chi^2 = \Sigma\frac{(O - E)^2}{E} = 3.8 + 3.8 = 7.6$$

poll may be only a chance occurrence. Thus, he will apply a χ^2. He has observed frequencies of 114 and 76 and will compare these frequencies against an expected 50:50 (95 and 95) distribution as seen in Table 10.2.

Note that Table 10.2 follows the important rule that the totals of the observed (190) must equal the totals of the expected (190). Jack now adds new entries based upon calculations between observed and expected and makes a final χ^2 determination as shown in Table 10.3

Since there are two categories (approve and disapprove), this design has $2 - 1 = 1$ degree of freedom (*df*). The value in the χ^2 table (Table 4 in Appendix D) for 1 *df* at the .05 level is 3.84. Since the resulting χ^2 value of 7.6 is greater than

Table 10.4.   Chi Square Determination for a 75:25 Distribution

	Approve	Disapprove	Total
O	114	76	190
E (75:25)	142.5	47.5	190
$O - E$	−28.5	+28.5	
$(O - E)^2$	812.25	812.25	
$\dfrac{(O-E)^2}{E}$	5.7	17.1	22.8

$$\chi^2 = \Sigma \frac{(O-E)^2}{E} = 5.7 + 17.1 = 22.8$$

3.84, Jack concludes that the sample is not a 50:50 distribution. He may now state that his poll indicates that more than 50% of the plant's workers would approve of the new design. If the resulting χ^2 had been less than 3.84, he would have concluded that the distribution was 50:50.

10.2.2 Example of Another Distribution. A distribution other than 50:50 may also be tested. Suppose Jack had decided that he would purchase new glasses only if more than 75% of all workers approved. This design also has 1 df as before. Table 10.4 shows the χ^2 determination. Since the determined value of 22.8 exceeds the value of 3.84 at the .05 level, Jack concludes that the sample is not 75:25 distributed and that less than 75% of all workers approve of the new design.

10.2.3 Example of a Distribution with More Categories. This technique may be applied to more than two groups or categories. A company's 300 VDT operators are polled to determine whether an amber, green, or black background for their monitor is most uncomfortable to use. The researcher believes that operators have no particular aversion to any of the three colors. The results are shown in Table 10.5.

Since there are three categories, $df = 3 - 1 = 2$. In the χ^2 table, 2 df at the .05 level has a value of 5.99. Since the final χ^2 value of .78 does not reach 5.99, the researcher concludes that the poll is evenly distributed at a 33.3:33.3:33.3 ratio and that the operators have no particular color aversion.

10.3 Multiple Comparisons. With the exception of section 7.3 (comparing three or more test results), all of our statistical procedures thus far have been

Table 10.5.   Aversion to Monitor Screen Background Color

	Amber	Black	Green	Total
O	105	102	93	300
E	100	100	100	300
(33.3:33.3:33.3)				
$O - E$	+5	+2	−7	
$(O - E)^2$	25	4	49	
$\dfrac{(O-E)^2}{E}$	.25	.04	.49	.78

$$\chi^2 = \Sigma \, \frac{(O - E)^2}{E} = .25 + .04 + .49 = .78$$

Table 10.6.   Running Time and Down Time for Three Ventilation Systems (in Hours)

	System 1	System 2	System 3	Totals
Running time	160	155	140	455
Down time	0	5	20	25
Totals	160	160	160	480

based upon a comparison between two groups. We will now describe χ^2 techniques for comparing three or more groups.

10.3.1 Example: Comparing Three Groups and Two Categories. Mary is an industrial hygienist who is involved with the ventilation systems of three chemical operations in her plant. She has been counting the number of failure hours and running hours of each system of the three operations. She strongly suspects that the ventilation system of Operation 3 is particularly ineffective and might need to be overhauled. She wishes to discover whether this system is different from the other two. From maintenance records, she has collected the information in Table 10.6 on the three systems for the last four 40-hour weeks.

Table 10.6 contains observed frequencies only. Expected frequencies must be calculated and appropriately placed into each of the six cells in the table. These are calculated by multiplying the column total and row total of a cell and then dividing by the grand total. As an example, the expected frequency

Table 10.7. Observed and Expected Frequencies

	System 1		System 2		System 3		Totals	
	O	*E*	*O*	*E*	*O*	*E*	*O*	*O*
Running time	①160	151.7	②155	151.7	③140	151.7	455	455.1
Down time	④ 0	8.3	⑤ 5	8.3	⑥ 20	8.3	25	24.9
Totals	160	160	160	160	160	160	480	480

Note: The circled numbers correspond to the numbers of the equations in the text.

for the cell "System 1/Running Time" is 160 × 455 divided by 480 equals 151.7. Table 10.7 shows the observed and expected frequencies. The final calculation for the χ^2 value is as follows:

$$\chi^2 = \Sigma \frac{(O - E)^2}{E}$$

$$\frac{(160 - 151.7)^2}{151.7} = .45 \tag{1}$$

$$\frac{(155 - 151.7)^2}{151.7} = .07 \tag{2}$$

$$\frac{(140 - 151.7)^2}{151.7} = .90 \tag{3}$$

$$\frac{(0 - 8.3)^2}{8.3} = 8.3 \tag{4}$$

$$\frac{(5 - 8.3)^2}{8.3} = 1.31 \tag{5}$$

$$\frac{(20 - 8.3)^2}{8.3} = 16.49 \tag{6}$$

$$\chi^2 = 27.52$$

The *df* for this design is 3 groups − 1 = 2 *df*. At the .05 level, 2 *df* = 5.99. Since the resulting χ^2 value of 27.52 exceeds 5.99, there is a difference between the three systems. Mary logically concludes that the difference lies in System 3 because of its large down time.

10.3.2 Example: Testing Multiple Groups with Multiple Categories. There is a design that may be used for more than two groups that shows how they relate to more than two categories. A student writing a master's thesis has, over the course of a year, examined 1,000 admissions records in the emergency room of a local hospital. The records describe workers who have experienced an industrial accident. Each record contains a notation about the worker's racial or ethnic group and a notation about the estimated severity of the accident injury: minor, major, or grave. The student wishes to determine whether belonging to any particular racial or ethnic group influences accident frequency and severity. His findings appear in Table 10.8.

To perform calculations, an expected frequency must be inserted into each cell. The expected frequency is obtained by multiplying the total of the column to which the cell belongs by the total of the row to which the cell belongs and then dividing by the grand total. As an example, in Table 10.9, the expected

Table 10.8. 1,000 Emergency Room Accident Records

	Caucasian	African American	Asian American	Totals
Minor	192	180	220	592
Major	83	90	85	258
Grave	62	48	40	150
Totals	337	318	345	1,000

Table 10.9. Observed and Expected Frequencies

	Caucasian		*African American*		*Asian American*		*Totals*	
	O	*E*	*O*	*E*	*O*	*E*	*O*	*E*
Minor	①192	199.5	②180	188.3	③220	204.2	592	592
Major	④ 83	87.0	⑤ 90	82.0	⑥ 85	89.0	258	258
Grave	⑦ 62	50.6	⑧ 48	47.7	⑨ 40	51.7	150	150
Totals	337	337.1	318	318.0	345	344.9	1,000	1,000

Note: The circled numbers correspond to the numbers of the equations in the text.

frequency of cell 1 is 337 × 592 divided by 1,000 = 199.5. χ^2 is then solved by a series of calculations similar to those performed in prior sections:

$$\chi^2 = \Sigma \frac{(O - E)^2}{E}$$

$$\frac{(192 - 199.5)^2}{199.5} = .28 \tag{1}$$

$$\frac{(180 - 188.3)^2}{188.3} = .37 \tag{2}$$

$$\frac{(220 - 204.2)^2}{204.2} = 1.22 \tag{3}$$

$$\frac{(83 - 87.0)^2}{87.0} = .18 \tag{4}$$

$$\frac{(90 - 82.0)^2}{82.0} = .78 \tag{5}$$

$$\frac{(85 - 89.0)^2}{89.0} = .18 \tag{6}$$

$$\frac{62 - 50.6)^2}{50.6} = 2.57 \tag{7}$$

$$\frac{(48 - 47.7)^2}{47.7} = .002 \tag{8}$$

$$\frac{(40 - 51.7)^2}{51.7} = 2.65 \tag{9}$$

$$\chi^2 = 8.23$$

The final χ^2 result is 8.23. It is now important to discover the degrees of freedom (df) of this design. Prior to studying this design, df was determined simply by subtracting 1 from the number of columns found in the design. Now, because there are more than two rows, the formula $(k - 1)(r - 1) = df$ must be applied. Applying this formula produces $(3 - 1) \times (3 - 1) = 4$ df. It should be noted that this formula applies in all forms of the χ^2 including the simple designs found earlier in this book. We now have the final result of 8.23 at 4 df. Checking the χ^2 table, we discover that, at the .05 level, 9.49 is required for significance. Since our value of 8.23 does not reach 9.49, we conclude that there is no difference in accident frequency and severity of the three groups and that no racial or ethnic group has a proclivity toward industrial trauma.

Problems

10.1 Ms. Adams is a human factors engineer who is designing a lever that will be used as a machine control. She is not sure whether it would be more effective to provide a 10-pound or a 15-pound resistance spring on the lever. She decides to experiment on the lever using both types of springs. Using the same operator over many trials for each spring, she counts the total of manipulating errors as well as how they are distributed. Table 10.10 presents her results.

   a. Apply a χ^2 using a 50:50 distribution.
   b. Which spring is the better choice?

10.2 Ms. Adams wishes to reconsider her treatment of the data in Table 10.10. She had also been keeping count of the total number of trials for each spring. These are shown in Table 10.11. Taking the number of trials into consideration:

   a. Apply a χ^2 such as that found in Chapters 3, 4, 5, and 8.
   b. Should Ms. Adams change her conclusion about the springs?

Table 10.10. Manipulating Errors

10-Pound Spring	15-Pound Spring	Total Errors
98	104	202

Table 10.11. Resistance Spring Trials

	10-Pound	15-Pound	Total
Trials with error	98	104	202
Trials without error	604	1,153	1,757

10.3 A safety officer has read that though older workers generally experience fewer total accidents than their younger counterparts, they may suffer a greater incidence of slips, trips, and falls. He examines all company accident and injury records for the last five years. He finds 203 reports related to slips, trips, and falls. Each report contains the injured worker's date of birth. The information appears in Table 10.12.

   a. Make a comparison against a 25:25:25:25 distribution.
   b. What is the conclusion about the older workers?

10.4 Ms. Rogers is a corporation nurse who is concerned that company employees are generally overweight. She wishes to start an Overweight Campaign and would like to support the campaign with some hard facts about the employees. For one week she weighed and measured all of the 62 employees who visited her station. Using standardized tables, she was able to determine how many of the 62 employees were overweight. The facts are presented in Table 10.13. Ms. Rogers intends to state in the company newsletter that currently 25% of company employees are overweight and should begin a safe weight reduction program. Is she justified in making this statement?

10.5 Is there such a thing as "Blue Monday"? A manufacturing company, which works a five-day week, has presented the data in Table 10.14 to a magazine writer at his request. The data show the number of visits to company first aid stations on each day of the week over the course

Table 10.12. Number of Workers Falling by Age Group

Age 18–29	Age 30–41	Age 42–53	Age 54–65	Total
47	50	49	57	203

Table 10.13. Sampling of Company Employees

Overweight	Acceptable Weight	Total
15	47	62

of a year. Would the writer be justified in submitting an article claiming that there is a "Blue Monday" for American industry?

10.6 Jackson is a loss control manager for an insurance company. He has four subordinates who are loss control inspectors. Their responsibilities are to perform safety inspections in local industrial plants. These inspectors are also required to make safety recommendations as a result of their inspection. Recommendations are classified as Desirable, Important, or Urgent. Jackson is concerned that Jim, one of the inspectors, is not doing a very conscientious job. Jackson analyzes 50 inspection reports of each inspector and summarizes his findings in Table 10.15. Does Jim need extra training and counseling?

Table 10.14. Visits to the First Aid Station

Monday	Tuesday	Wednesday	Thursday	Friday	Total
69	45	50	46	50	260

Table 10.15. Recommendations of Three Loss Control Inspectors

	Diane	John	Jim	Total
Desirable	102	105	91	298
Important	35	37	27	99
Urgent	10	11	5	26
Total	147	153	123	423

Chapter 11

Identifying Accident Repetition

11.1 Background. Accident prone, accident proneness, accident repeater, and accident repetition are controversial terms that are recognizable to both the safety professional and the general public. There is substantial disagreement over whether accident repetition is a real phenomenon and over whether accident repeaters actually exist. This chapter will not present either a rationale for or an argument against this concept. Instead this chapter was written with the intent of providing a research technique that might aid those professionals who support the theory of accident repetition. First, though, to provide some information about this area, some thoughts and questions are posed.

11.1.1 Accident Proneness versus Accident Repetition. What are the differences between *accident proneness* and *accident repetition?* Both concepts involve the frequent occurrence of incidents or accidents to one individual within a relatively short period of time. Studies based upon these concepts are usually devoted to studying frequencies rather than severities. Accident proneness and accident repetition are subtly dissimilar, however. Although opinions differ, proneness is best defined as a substantial frequency of occurrence due to some ostensible physical limitation such as visual, hearing, mental, or neurological impairment. Belonging to a group, such as males under the age of 25, implies proneness to some people, but this use of the term is questionable because it encompasses an entire group rather than an individual. Repetition also involves substantial frequency, but no physical or psychological explanation is currently apparent, although such an explanation may be discovered later after the repeater is identified.

11.1.2 The Number of Accident Repeaters. Are there many accident repeaters? Should such people exist, they would be very few, perhaps a fraction of 1% of a given population. Those who support the repetition theory and those who support procedures for identifying repeaters argue that although repeaters are a small segment of the population, their tendency toward accidents will eventually lead to severe injuries that are costly to them and to their organization. If, and when, such repeaters are identified, a managerial and philosophical dilemma often arises about how to protect the organization and the repeater from him or herself. Discharging repeaters or limiting them to certain activities may lead to negative organizational or legal repercussions. Allowing them to function in their present circumstances may be ethically unpardonable.

11.1.3 The Permanence of Accident Repetition. Is accident repetition a permanent condition? Those who support the repetition concept admit that repeaters do not necessarily remain repeaters. Repeaters may alter their patterns of accident frequency because of psychological and physical changes within themselves or social, managerial, or environmental changes external to themselves. An individual considered a repeater in one occupation may not be a repeater in another occupation even though the new occupation may be considered equally as hazardous as the first. It is also possible that an individual was incorrectly labeled as a repeater originally and has subsequently displayed an improved performance that approximates those of other individuals.

11.2 The Basic Requirements for Identifying a Repeater. If a researcher believes that he or she can succeed in identifying a repeater, each of the following required characteristics must be considered.

11.2.1 Absence of Obvious Physical or Psychological Impairments. The suspected individual should have no obvious physical or psychological characteristics that would with certainty make her or him a repeater. As an example, an individual with very limited vision would almost certainly be a repeater on a construction site. In all likelihood, though, such an individual would not have been selected for the job to begin with. Only suspected individuals with ambiguous characteristics or no characteristics at all (except for their accident frequency record) would be an object of this research.

11.2.2 High Frequency of Accidents. The suspected repeater should currently display a substantially high frequency of accidents or incidents compared to his or her fellow workers. The cutoff point between a high frequency record and an average record is a matter of judgment left to the researcher. It is recommended that the cutoff be at a high enough level so that only those who are highly suspect are investigated. With a low cutoff, a great deal of time and expense may be expended for only minimally satisfactory results.

11.2.3 Consistency and Regularity of Accidents over Time. The suspected repeater must not only manifest a high frequency of accidents, but he or she should experience these accidents consistently and regularly over the course of time. It is not appropriate to label an individual a repeater based upon a high accident frequency for only one accident recording period. This circumstance is analogous to throwing a die over a specified number of runs and having the number 6 occur every time. This outcome is improbable but still possible and a part of chance variation. If, however, the die is thrown again over the same number of specified runs and 6 again occurs all of the time, then we would suspect that something is wrong. In the case of the die, we might determine that it is not a perfectly symmetrical cube. In the case of a human being who has a high accident frequency from one recording period to the next, we might determine that there is something unusual about him or her that causes accident repetition.

11.3 Procedure and Example. A recommended method for identifying accident repeaters involves the following steps.

11.3.1 List Employees and Reported Accidents. From prior accident recording periods, list those employees who were working full-time during these periods and who experienced at least one reportable accident. It does not matter if the periods are years, half-years, quarters, or some other unit; however, they must be contiguous and equal in length. Record the number of reported accidents for these employees for each period as in Table 11.1.

Table 11.1. Number of Reported Accidents

	1994		1995		1996		
	First Half	Second Half	First Half	Second Half	First Half	Second Half	Total
Joe	2	2	3	1	3	2	13
Megan	1	2	0	0	3	0	6
Charles	0	0	0	1	0	0	1
John	0	0	1	0	0	2	3
Enrique	0	1	0	0	0	0	1
Mike	1	2	0	0	1	0	4
Nora	0	7	0	1	1	0	9
Jason	0	0	0	1	0	0	1
Sam	0	0	2	2	1	2	7
							45

11.3.2 Rank Employees and Find the Median. Next rank the employees according to their overall totals and identify the median as in Table 11.2.

11.3.3 Perform Chi Square Computations. Working with those totals higher than the median, and starting with the highest total, begin a series of chi square computations comparing each total against a .333:.666 distribution. These computations are shown in Table 11.3.

Table 11.2. Reported Accidents

Charles	1	
Enrique	1	
Jason	1	
John	3	
Mike	4	(Median)
Megan	6	
Sam	7	
Nora	9	
Joe	13	
	45	

Table 11.3. Chi Square Computations

	Reported Accidents		
	Joe	**Other Employees with Accidents**	**Total**
Observed (*O*)	13	32	45
Expected (*E*) (.333:.666)	15	30	45
O − *E*	−2	+2	
(*O* − *E*)^2	4	4	
$\dfrac{(O - E)^2}{E}$	.27	.13	

$$\chi^2 = \Sigma \frac{(O - E)^2}{E} = .27 + .13 = .40$$

Table 11.3   Chi Square Computations—*Continued*

| | | *Reported Accidents* | |
	Nora	Other Employees with Accidents	Total
Observed (O)	9	36	45
Expected (E) (.333:.666)	15	30	45
$O - E$	−6	+6	
$(O - E)^2$	36	36	
$\dfrac{(O - E)^2}{E}$	2.4	1.2	

$$\chi^2 = \Sigma \frac{(O - E)^2}{E} = 2.40 + 1.20 = 3.60$$

| | | *Reported Accidents* | |
	Sam	Other Employees with Accidents	Total
Observed (O)	7	38	45
Expected (E) (.333:.666)	15	30	45
$O - E$	−8	+8	
$(O - E)^2$	64	64	
$\dfrac{(O - E)^2}{E}$	4.27	2.13	

$$\chi^2 = \Sigma \frac{(O - E)^2}{E} = 4.27 + 2.13 = 6.40$$

11.3.4 Continue Calculations.   Calculations continue until a resulting value (Sam) exceeds 3.84 (χ^2 of 1 *df* at .05 level). In this study, Joe (χ^2 of .40) and Nora (χ^2 of 3.6) do not reach 3.84, and it is concluded that Joe and Nora may each comprise 33.33% of all accidents experienced by the group. Of the workers in the group, Joe and Nora are now considered candidates for being called accident repeaters because of their high proportion of accidents within the entire group. High frequency is not enough, however. For Joe and Nora to be considered repeaters, their accident experiences must be proved to be consistent.

11.3.5 Analyze Subtotals. Joe's and Nora's subtotals for each period are now analyzed with a χ^2 test against a distribution of 16.6:16.6:16.6:16.6: 16.6:16.6. The computations appear in Table 11.4.

Table 11.4. Analysis of Joe's and Nora's Subtotals

| | Joe's Accidents over Three Years | | | | | | |
| | 1994 | | 1995 | | 1996 | | |
	First Half	Second Half	First Half	Second Half	First Half	Second Half	Total
Observed (O)	2	2	3	1	3	2	13
Expected (E) (.166)	2.16	2.16	2.16	2.16	2.16	2.16	12.96
$O - E$	−.16	−.16	+.84	−1.16	+.84	−.16	
$(O - E)^2$	.03	.03	.70	1.35	.70	.03	
$\dfrac{(O - E)^2}{E}$	.01	.01	.32	.63	.32	.01	

$$\chi^2 = \Sigma \frac{(O - E)^2}{E} = .01 + .01 + .32 + .63 + .32 + .01 = 1.30$$

| | Nora's Accidents over Three Years | | | | | | |
| | 1994 | | 1995 | | 1996 | | |
	First Half	Second Half	First Half	Second Half	First Half	Second Half	Total
Observed (O)	0	7	0	1	1	0	9
Expected (E) (.166)	1.499	1.499	1.499	1.499	1.499	1.499	8.99
$O - E$	−1.499	5.5	−1.499	−.499	−.499	−1.499	
$(O - E)^2$	2.25	30.25	2.25	.25	.25	2.25	
$\dfrac{(O - E)^2}{E}$	1.5	20.18	1.5	.17	.17	1.5	

$$\chi^2 = \Sigma \frac{(O - E)^2}{E} = 1.5 + 20.18 + 1.5 + .17 + .17 + 1.5 = 25.02$$

11.3.6 Identify Possible Repeaters. The df of this design equals $(6 - 1) \times (2 - 1) = 5$ df. The χ^2 value of 5 df at the .05 level equals 11.07. Nora, with a χ^2 value of 25.02 exceeds this value; thus, it is concluded that Nora's accidents are not evenly dispersed over the six periods. Joe's χ^2 value of 1.30 does not exceed 11.07; thus, it is concluded that Joe's accidents are evenly dispersed over the six periods. The ultimate conclusion is that there is one employee to be concerned about and that employee is Joe because (1) he has a substantial number of total accidents compared to other employees and (2) his accident behavior is consistent and may very well continue into the future.

11.4 Important Notes. Four important points should be made: (1) This present method used a .333:.666 distribution as a criterion to identify an excessively high accident rate for each worker. A researcher may decide on another distribution such as 25:75 or 50:50. This decision is a matter of judgment; however, the distribution should be stated before research begins and should be used consistently thereafter. (2) Calculations are probably not necessary for totals below the median because no significant conclusions are likely to result. (3) Accident repeaters do not necessarily turn up every time these calculations are made. (4) Finally, this method would not readily succeed on a small population of workers, especially one that displays infrequent accident occurrences.

Problems

11.1 Using your own distribution criterion, analyze the data in Table 11.5 to discover whether there may be any accident repeaters.

Table 11.5. Number of Reported Accidents over Eight Quarters

	1995					1996			
	1Q	2Q	3Q	4Q		1Q	2Q	3Q	4Q
Mancuso	2	0	0	0		1	0	0	0
Tanaka	1	1	0	3		0	0	0	0
O'Brien	3	0	3	1		3	3	1	0
Jackson	4	5	6	0		0	0	0	0
Higgins	2	0	2	0		0	2	0	0
Murray	1	0	0	0		0	0	0	0
Vasquez	2	1	2	0		2	1	2	0
Hamilton	6	2	2	0		0	1	0	0
Ransom	0	0	0	0		0	0	0	1
Johnson	0	0	3	0		4	1	0	0

11.2 Using a .33:.66 distribution, analyze the data in Table 11.6 to discover whether there may be any accident repeaters.

Table 11.6. Recorded Accidents over Twelve Months

Employee #	Jan.	Feb.	Mar.	Apr.	May	June	July	Aug.	Sept.	Oct.	Nov.	Dec.
138	0	1	0	0	0	0	0	0	0	0	0	0
184	0	0	0	0	1	0	0	0	0	0	0	0
95	1	0	0	0	0	0	1	0	0	0	0	0
86	0	0	0	0	0	0	0	0	0	1	0	0
201	0	0	0	0	2	1	0	0	0	0	0	0
230	0	1	0	1	0	1	0	1	0	1	0	0

Chapter *12*

Administrative Controls

12.1 Discussion. It is important that an organization maintain control over all its internal functions and processes including safety and health. These controls are essentially monitoring procedures that test the organization's "status quo" to make certain that any systems that have been installed are operating satisfactorily. Many of the procedures described so far in this book may be considered control procedures, especially those in Chapter 8 dealing with frequency and severity rates. This chapter offers two additional control methodologies that may be of practical use to the safety and health professional.

12.2 Control Charts. *Control charts* are control instruments based upon a statistical concept known as *confidence intervals.* In safety and health, these charts are most frequently applied to accident frequencies, but they may also be applied to many other variables as well. Basically, a control chart reports the positive or negative changes in magnitude of a particular variable over successive, contiguous periods of time. The chart also reports when a change of magnitude for the latest (present) period has reached such a positive or negative extreme that the safety professional should pay some extra attention to the current safety situation.

12.2.1 Procedure and Example. In this example, which is typical of this procedure, accident frequency rates will be placed in a control chart. The frequency rates may already have been computed; if not, it is advisable to perform this computation in order to take the organization's "busyness" into consideration. It would also be acceptable to insert accidents per full-time worker into the chart or just the number of accidents themselves, provided that

Table 12.1.   Johnson Corporation: Quarterly Reportable Accidents, 1991–1994

	Number of Accidents	Worker Hours	Accident Rate Using 200,000 (X)	Rate Squared (X^2)
1991	13	200,502	12.97	168.22
	16	201,145	15.91	253.13
	12	190,888	12.57	158.00
	12	204,000	11.76	138.30
1992	9	199,306	9.03	81.54
	10	202,188	9.89	97.81
	14	210,005	13.33	177.69
	13	200,780	12.95	167.70
1993	17	202,123	16.82	282.91
	14	200,101	13.99	195.72
	11	199,853	11.01	121.22
	8	190,900	8.38	70.22
1994	10	191,633	10.44	108.99
	16	201,500	15.88	252.17
	12	206,974	11.60	134.56
	13	201,701	12.89	166.15
			199.42	2,574.33

the number of full-time workers in the organization or its "busyness" has not changed. Table 12.1 lists accident occurrences for a large corporation over four years in its first column and provides calculations needed for the chart in the other columns.

In the table, each item in the first column (number of accidents) is divided by the item in the second column (worker hours). This number is then multiplied by a large quantity (in this case, 200,000) to produce a manageable number, which is also the frequency rate. This frequency rate for the period in question is placed in the third column. The frequency rate is then squared, and the result is placed in the fourth column. Next, the third and fourth columns are added up. The sum of the frequencies (199.42) is then divided by the number of periods (16) to obtain an average frequency of 12.46. Other calculations follow. Note that X is the symbol for frequency rates, X^2 is the symbol for frequency rates squared, and $\overline{X}$ is the symbol for average frequency rate.

$$\text{Standard error (SE)} = \frac{\sqrt{\dfrac{\Sigma X^2}{N} - \overline{X}^2}}{\sqrt{N - 1}}$$

$$SE = \frac{\sqrt{\dfrac{2{,}574.33}{16} - 12.46^2}}{\sqrt{16 - 1}}$$

$$SE = \frac{\sqrt{160.90 - 155.25}}{\sqrt{15}} = \frac{2.38}{3.87} = .62$$

$$df = N - 1 = 16 - 1 = 15$$

t value at 15 df at .01 level $= 2.947$

$$\text{Control limits} = \overline{X} \pm t \text{ value (SE)}$$

$$\text{Upper control limit (ULC)} = \overline{X} + 2.947(\text{SE})$$

$$ULC = 12.46 + 2.947(.62) = 14.29$$

$$\text{Lower control limit (LCL)} = \overline{X} - 2.947(\text{SE})$$

$$LCL = 12.46 - 2.947(.62) = 10.63$$

This calculation based upon information from Table 12.1 provides a result known as the standard error (SE), which has already been discussed in Chapter 8. This number will be the basis for determining the control limits. To do this, reference must be made to Table 2 in Appendix D. The table is referenced first by df, which in this design is $N - 1 = 16 - 1 = 15$. Working at a .01 level of probability, the value opposite the df 15 row at the .01 column is 2.947. The

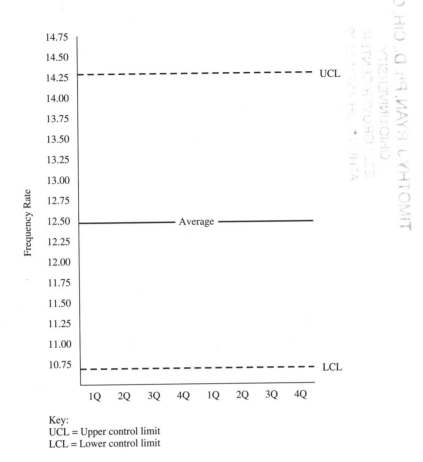

Key:
UCL = Upper control limit
LCL = Lower control limit

Figure 12.1.   Control Chart Layout

upper control limit is equal to the average frequency rate plus 2.947 multiplied by the standard error. Likewise, the lower control limit is equal to the average frequency rate minus 2.947 multiplied by the standard error. The control limit values of 14.29 and 10.63 are then placed upon a chart as lines, along with a line representing the average frequency rate (12.46). Figure 12.1 shows an example of how a control chart may be constructed for this particular case.

Note three important considerations: (1) The *df* of this type of design always varies with the *N*, which is the number of time intervals. To make proper use of the table, the *df* must be determined carefully. (2) The larger the number of time intervals used, the more reliable the study will be. (3) The .05 level of probability is also available from Table D.2 in Appendix D. This level is a more conservative approach that narrows the band between the control limits.

Table 12.2. Johnson Corporation:
Accident Frequency Rates, 1995–1996

1995	12.98
	13.75
	12.25
	11.44
1996	13.79
	14.10
	14.60
	14.75

12.2.2 Using the Control Chart. Having established a control chart based upon frequency rates that occurred in the immediate past, we will now follow Johnson Corporation's progress into the future. Table 12.2 shows new quarterly frequency rates calculated upon accidents and work hours that occurred since the chart was constructed.

Figure 12.2 shows these rates displayed upon the previously constructed control chart. Observe that the second half of 1996 experienced rates that were above the upper control limit. This would indicate to the Safety Department that some unusual circumstances have developed, physically and/or managerially, within the organization, which have led to and probably will continue to lead to an undesirable accident rate. Unfortunately, the chart tells us only that something is wrong, not what is wrong. Finding out what is wrong will require further research. It should also be noted that if any rates had fallen below the lower control limit, then something good is going on. The Safety Department should find out what it is and keep applying it to the organization. Finally, it should be emphasized that the company's control chart should be periodically updated by using the most recent data. In this particular case, a new chart should be recalculated at approximately two-year intervals.

12.3 Cost Controls. In addition to ethical and humanitarian concerns, safety professionals agree that one of the major reasons for applying safety systems to an organization is to diminish costs and enhance profitability. Safety management textbooks always address the importance of communicating to management the overt and hidden costs of an accident or a disease. These books do not offer any established or standardized quantified methods for computing these costs, however. The main reason for this is that costs and cost accounting procedures vary considerably between companies because their individual organizations and financial philosophies differ significantly from one another.

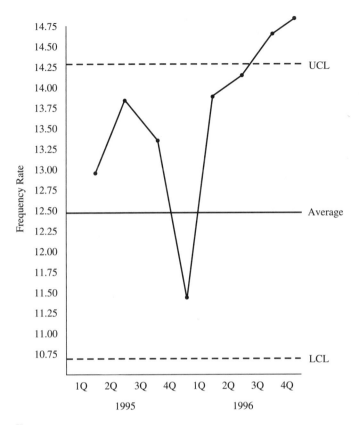

Key:
UCL = Upper control limit
LCL = Lower control limit

Figure 12.2. Johnson Company Accident Frequency Rate

12.3.1 Safety Expenditure Control. Although there is no standardized method of accident cost accounting, there is at least one reasonably standardized method for controlling costs related to spending money on safety improvements. This method takes into consideration all types of safety expenditures, whether physical (such as installing sprinkler systems, upgrading electrical systems, installing new machine guards, and the like) or managerial (such as instituting new safety training programs, hiring more safety personnel, writing a new safety program, and the like). This method basically answers the question of whether the money spent on the improvement is financially worthwhile and whether the money saved by diminishing accidents and disease costs is substantially larger than the original safety expenditure.

12.3.2 Case Study: The Present Worth of Money Technique. A fire safety officer and the treasurer of a corporation are working together as a team to determine the financial feasibility of installing an expensive sprinkler system in one of their buildings. After researching some sources of information within and outside the corporation, they determine the following essential facts, which are needed to form an opinion:

1. Cost of installation by a competent contractor: $350,000
2. Life expectancy of the building and thus of the sprinkler system: 30 years
3. Annual savings in fire losses and fire insurance premiums: $25,000
4. Rate of interest return the corporation expects from investing its excess monetary assets: 8%

This information is inserted into the following calculation:

Symbols: Proposed expenditure (PE) = 250,000

Life expectancy (L) = 30

Annual savings (S) = 25,000

Value of money (V) = .08

Present worth of money (PW_M)

Present worth of savings (PW_S)

Formulas: $$PW_M = \frac{(1 + V)^L - 1}{V(1 + V)^L}$$

$$PW_S = PW_M(S)$$

Computation: $$PW_M = \frac{(1 + .08)^{30} - 1}{.08(1 + .08)^{30}}$$

$$PW_M = \frac{9.06}{.81} = 11.18$$

$$PW_S = 11.18(25,000) = \$279,500$$

$$PE = \$250,000$$

$$PW_S \text{ (of } \$279,500) > PE \text{ (of } \$250,000)$$

Since the present worth of savings is larger than the proposed expenditure, it is financially advantageous and feasible for the corporation to install the sprinklers. If the present worth of savings were substantially less than the proposed expenditure, corporation management might have reservations about investing in the sprinklers. Nevertheless, despite the financial disadvantage, the sprinklers might still be installed for the humanitarian purpose of guaranteeing life safety to employees. It is advisable that a procedure such as this be applied in situations where there may be large safety expenditures. The one apparent limitation in applying this procedure is the question of subjectivity. Methods of predicting life expectancy and estimating annual savings involve a good deal of judgment that is only partially tempered with limited amounts of research.

Problems

12.1 A corporation's monthly accident frequencies for 1993 and 1994 are reported in Table 12.3.

   a. Based upon these data, construct a control chart with upper and lower control limits based upon the .05 or .01 level of probability. Base the frequency rate upon 200,000.
   b. The frequencies and hours worked for 1995 appear in Table 12.4. Plot them on the control chart and state a conclusion.

Table 12.3. Monthly Accident Frequencies

1993	Frequencies	Worker Hours	1994	Frequencies	Worker Hours
January	2	12,890	January	4	13,100
February	1	13,301	February	3	13,182
March	3	10,983	March	0	9,964
April	4	10,959	April	0	11,158
May	2	13,101	May	1	10,999
June	2	12,464	June	3	12,656
July	3	12,737	July	2	12,509
August	4	10,103	August	2	11,994
September	2	12,499	September	3	12,476
October	1	13,460	October	3	12,664
November	0	12,804	November	2	13,001
December	2	12,002	December	2	12,503

Table 12.4. Accident Frequencies, 1995

	Frequencies	Hours
January	2	12,603
February	2	11,138
March	1	10,144
April	3	10,880
May	6	9,111
June	7	9,125

Table 12.5. Number of Pilot's Errors: 30 Pilots for One Hour

Pilot Number	Errors	Pilot Number	Errors	Pilot Number	Errors
1	4	11	2	21	5
2	6	12	3	22	1
3	2	13	4	23	3
4	1	14	6	24	3
5	3	15	1	25	4
6	5	16	1	26	3
7	4	17	2	27	2
8	3	18	3	28	2
9	2	19	1	29	3
10	1	20	2	30	1

12.2 Captain Ross is a human factors engineer who is interested in pilot error as related to pilot fatigue. He has been testing pilot trainees in a flight simulator for one-hour periods. During the one-hour simulation, the simulator has been recording each pilot trainee's operating errors. These are recorded in Table 12.5.

   a. Construct a control chart reflecting the performance of the 30 pilots. Include control limits based upon the .05 level of probability.
   b. Captain Ross now continues his study. He asks each of the 30 pilots to use the simulator for four straight hours and counts their

Table 12.6. Average Number of
Pilot Errors: Half-Hour Intervals

Interval	Average Number of Errors
1	2.4
2	2.9
3	3.2
4	4.3
5	4.9
6	5.4
7	5.9
8	6.2

Table 12.7. Safety Rule Violations Observed, 1995

Month	Count	Month	Count
January	38	July	40
February	45	August	39
March	42	September	35
April	35	October	35
May	33	November	36
June	37	December	32

errors for each progressive half-hour. He then averages these errors for each progressive half-hour. The data are shown in Table 12.6. Record these data on the control chart. At what point does performance deteriorate because of fatigue?

12.3 John is a safety officer for a construction company. He believes that an indicator of correct safety attitude is worker observance of safety rules. John periodically sends out safety inspectors to construction sites to observe workers' behaviors. Approximately the same number of workers are observed each month. Table 12.7 shows the monthly number of safety rule violations for year 1995.

    a. Using the .01 level of probability, construct a control chart for the data.

    b. Additional observation results for the first half of 1996 appear in Table 12.8. What can be concluded?

12.4 A corporation with a small Safety Department and a mediocre safety record is considering hiring a highly respected safety consulting firm for a one-time fee of $100,000. The consulting firm says that it will install a safety program that is self-perpetuating for five years and will save the company $40,000 in accident and insurance costs annually. The company believes that a fair rate of monetary return is 10%. Would this project be financially advantageous to the company?

12.5 A company has been experiencing many worker's compensation claims because of employees slipping and falling. Management is seriously considering purchasing a new floor wax that substantially decreases the coefficient of friction of walking surfaces. The company believes that for an annual increase of $25,000 in floor wax expenditures it can cut down on slip and fall claims by $20,000 annually. The company believes that a fair rate of monetary return is 8%. Would this expenditure be advantageous to the company?

Table 12.8. Safety Rule Violations Observed, 1996

January	30
February	28
March	25
April	25
May	23
June	21

Chapter *13*

Time Series

13.1 Background. In Chapters 4 and 5, we discussed the concept of time as an important ingredient in the application of research. In this chapter, we will continue to use time as a tool in fact finding and problem solving. Our use of time in this respect satisfies our interest in history and provides knowledge about events in the past that may help us, by example, to solve our problems in the present and in the future.

13.2 Description. *Time series* is the study of changes, usually over a relatively long period of time, which are observed and recorded at smaller, equally spaced intervals of time. Table 13.1 presents data describing the death rates per 100,000 American children for the years 1913 to 1927. Although some conclusions and generalizations may be taken directly from these numbers, a graphic analysis would substantially improve our understanding of changes related to these data over time.

13.3 Graph Construction. The graph in Figure 13.1 (on p. 109) is initially constructed by drawing a horizontal straight line (*x*-axis) that meets a vertical straight line (*y*-axis) at a right angle. Traditionally, the *x*-axis represents time. This axis is broken down into smaller, equal intervals of time such as years, quarters, months, weeks, days, hours, and the like. Whether the intervals chosen are short or long depends upon the nature of the data. The physical spaces between these intervals along the *x*-axis will depend upon the researcher's judgment and preferences in the appearance of the graph. It is important to re-member two points: (1) The *x*-axis must contain only one unit of time. Thus, for example, weeks must not be mixed with months or days. (2) The physical intervals between time units must be of uniform width. The remaining axis (*y*-axis)

Table 13.1. Death Rates per 100,000 Children

Year	Age under Five	Age Five to Fourteen
1913	88.4	37.4
1914	94.3	38.9
1915	90.8	39.7
1916	101.4	43.3
1917	108.4	45.3
1918	91.0	46.5
1919	87.2	45.9
1920	87.4	44.9
1921	80.8	42.4
1922	80.6	41.5
1923	82.0	42.4
1924	82.9	42.4
1925	78.6	42.3
1926	77.9	41.4
1927	75.9	41.0

Source: National Safety Council.

represents the variable being studied such as deaths, injuries, or accidents, or dollars spent, earned, or lost. The researcher may choose many different variables depending on the research question at hand. The two rules about the *x*-axis are also approximately applicable to the *y*-axis; that is, (1) there can be only one unit of measurement upon the *y*-axis, and (2) the physical spaces between the intervals must be equal. One other rule should be remembered in setting up the axes. Time intervals on the *x*-axis move from earlier intervals on the left to later intervals on the right. Magnitudes of measurement on the *y*-axis move from smaller magnitudes at the bottom to greater magnitudes at the top.

13.3.1 Plotting. *Plotting* is the placement of a dot upon the graph at the intersection of the *y*-axis magnitude and the corresponding *x*-axis time interval. In plotting the graph in Figure 13.1, which is related to children under age five, the first plot was a dot placed opposite 88.4 (*y*-axis) and 1913 (*x*-axis). When all values are plotted, the dots are connected by straight lines. The lines may be continuous or composed of dashes or dots. After plotting is finished, both axes should be labeled and a concise comprehensible title given to the whole graph.

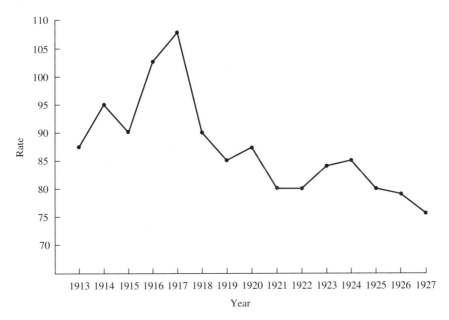

Figure 13.1.   Death Rates per 100,000 Children Age Five and Under

13.3.2 Interpretation.   Interpretations and conclusions are now made upon the configuration of the line on the graph. A researcher inspecting the graph in the year 1927 would probably conclude that deaths of young children were on the increase from 1913 to 1917, but then dropped substantially up to the present (1927). The line on the graph may go up, down, or stay level. Any prolonged movement in one direction is referred to as a *trend,* and this entire technique is referred to as *trend analysis.*

13.3.3 Multiple Variables.   It is possible to plot more than one variable upon a graph for purposes of comparison. When more than one variable is presented, all the variables must have some logical relationship to each other and be measurable by the units of measurement on the *y*-axis. Figure 13.2 adds a second variable (deaths of children aged 5 to 14) to the graph in Figure 13.1. A researcher inspecting this graph would likely conclude that (1) the death rate of older children is generally lower than that of younger children, but that (2) the death rate of older children is increasing slightly while the death rate of younger children is on the decline.

13.4 Trend Lines.   Because of confusing patterns and possible researcher bias, the trends represented by the data on a graph may be difficult to describe and interpret objectively. There is a method, however, that defines the trend of a series of graphic data both mathematically and objectively. This method

Figure 13.2. Death Rates per 100,000 Children

establishes a single trend line upon the graph that (1) represents all of the data quantitatively over time and (2) indicates an upward, downward, or horizontal movement. This method is referred to as the *least squares technique,* and its application is illustrated in the following example.

13.4.1 Example. Mr. Swenson is the newly appointed safety director of a large nationwide corporation. Upon taking the position, he has heard from various sources that the corporation's safety record has been deteriorating badly over past years. He feels it is important to verify this fact and decides to conduct some research. He is able to acquire from all corporate divisions data related to annual reported injuries and illnesses and annual numbers of full-time production workers. He then creates Table 13.2, which shows composite injury and illness incidents per 1,000 production workers for the entire corporation.

Table 13.2. Injuries and Illnesses
per 1,000 Production Workers

Year	Injuries and Illnesses
1984	88
1985	89
1986	90
1987	87
1988	92
1989	92
1990	93
1991	98
1992	100
1993	101
1994	105

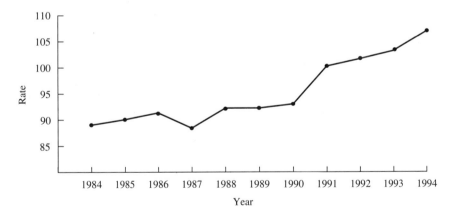

Figure 13.3. Injuries and Illnesses per 1,000 Production Workers

   Mr. Swenson plots these data as shown in Figure 13.3. He is able to identify
a steady increase in injury and illness incidents over the years, but would like to
provide an objective mathematical representation of these data for an important
report he will make to his board of directors. He constructs a trend line as follows.
First, he creates Table 13.3. The first column of this table contains the year.

Table 13.3. Calculations for Constructing a Trend Line

Year	X	Y	XY	X^2
1984	−5	88	−440	25
1985	−4	89	−356	16
1986	−3	90	−270	9
1987	−2	87	−174	4
1988	−1	92	−92	1
1989	0	92	0	0
1990	+1	93	+93	1
1991	+2	98	+196	4
1992	+3	100	+300	9
1993	+4	101	+404	16
1994	+5	105	+525	25
		1,035	+1,518	110
			−1,332	
			+186	

The second column, labeled X, contains an assignment of negative and positive values of 1 ascending or descending from the center-point year, which is assigned a zero. In this particular case, the center-point year is 1989.* The third column, labeled Y, is the injury and illness frequency for that year. The fourth column, labeled XY, is the algebraic product of columns X and Y. The fifth column, labeled X^2, is the square of column X. Columns X, XY, and X^2 are totaled. Note that XY is totaled algebraically by taking the negative values into consideration.

---

*In the event that there is an even number of years, assign an imaginary zero between the two years at the center; then use negative or positive, ascending or descending values of 1.0 starting from 0.5. If our data included the years 1984 to 1993:

1984	−4.5		1989	+0.5
1985	−3.5		1990	+1.5
1986	−2.5		1991	+2.5
1987	−1.5		1992	+3.5
1988	−0.5		1993	+4.5

The key formula for determining points on the trend line is Y(year) $= a + bX$. Y(year) is the Y value for a particular year, a is a constant base value, b is a constant multiplying factor, and X is the value assigned to a year. By using the information in Table 13.3, we may arrive at a and b:

$$a = \frac{\Sigma Y}{N} = \frac{1,035}{11} = 94.1$$

$$b = \frac{\Sigma XY}{\Sigma X^2} = \frac{186}{110} = 1.7$$

To construct the trend line, we use the values of a and b in our key formula. We do this once for the earliest year 1984, whose assigned X is -5, and once for the latest year 1994, whose assigned X is $+5$:

$$Y\,(\text{year}) = a + bX$$

$$Y\,(1984) = 94.1 + 1.7(-5) = 94.1 - 8.5 = 85.6$$

$$Y\,(1994) = 94.1 + 1.7(5) = 94.1 + 8.5 = 102.6$$

We now know that our trend line value for 1984 is 85.6 and our trend line value for 1994 is 102.6. We plot a Y value dot of 85.6 for 1984 and a Y value dot of 102.6 for 1994 and connect the dots to form the trend line. This line appears in Figure 13.4.

13.4.2 Interpretation and Prediction. Viewing the trend line, we can more confidently state that the safety record of the corporation is deteriorating as indicated by the upward inclination of the trend line. The steepness of the trend line also gives some indication about the rapidity of the deterioration. Finally, the trend line provides us with some limited ability for prediction. By extending the trend line on the graph into the next year (1995), we will have a prediction for that year. Any prediction from a trend line must be guarded, however, and must assume that there will be no major changes in the organization in question over the coming year.

13.4.3 Visual Fit. As previously stated, trends may also be described subjectively and nonmathematically by simply visually fitting a straight line upon the data according to judgment. This method, though practiced, is not as acceptable as the least squares technique.

13.5 Seasonality. *Seasons* are regular, dependable periodic fluctuations over time. Seasons are generally defined over the course of a year but may also

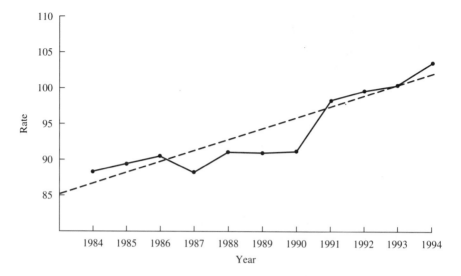

Figure 13.4.   Injuries and Illnesses per 1,000 Production Workers

be discovered over the course of a month or a week or a day. Many well-known examples of seasonality come to mind such as department store sales, which peak in December, and public accountants' activities, which peak in April. The study of seasons may help the safety and health professional understand patterns of occurrences within his or her organization that will lead to the control of injury and illness.

   13.5.1 Example.   Gina is the supervisor of a large department of television cable installers who work year round, mostly outdoors in Minnesota, and mostly at ground level. The installers experience a substantial number of slips and falls, and Gina believes that trauma from these falls occurs according to a well-defined pattern. She is able to collect data for a period of four years, as shown in Table 13.4. The size of the department and the number of hours worked have varied very little over these four years. The frequencies are listed on a monthly basis for each year, and an average monthly frequency for each year is calculated for future computations.

   Gina's next step is to perform a series of calculations in which each month's individual frequency is divided by the monthly average for the year. Thus, for example, the frequency of January 1993 is divided by the monthly average of 1993, or 8 divided by 5.4 equals 1.482. Likewise the frequency of January 1996 is divided by the monthly average of 1996, that is, 7 divided by 5.8 equals 1.207. All of these individual calculations appear in Table 13.5.

Table 13.4. Monthly Frequencies of Trauma due to Slips and Falls

Month	1993	1994	1995	1996
January	8	6	9	7
February	6	7	7	5
March	6	5	5	7
April	5	4	3	6
May	4	3	4	4
June	3	3	2	4
July	4	3	3	3
August	4	4	5	5
September	5	7	4	6
October	5	6	6	7
November	7	8	6	7
December	8	9	8	8
Totals	65	65	62	69
Monthly average	5.4	5.4	5.2	5.8

This process requires a substantial amount of arithmetic, so computational errors are possible. To check for possible errors, totals should be taken for the columns as well as for the rows. The total of the totals of the yearly columns should equal exactly, or almost exactly, the total of the totals of the monthly computations. Thus, $12.039 + 12.039 + 11.925 + 11.898 = 47.901$. The column on the extreme right is an average of the four monthly computations, or the total of the monthly entries divided by four. This column is the seasonal index for each month of the year based upon averaging all four years. To check for computational error, this column should be exactly, or almost exactly, equal to 12. If the column total is not exactly 12 and the researcher wishes to be very exact in reporting quantities in this column, then each quantity in the column should be adjusted by multiplying the quantity by the proportion of the column total and 12. In this case, the adjustment factor would be 12 divided by 11.979 or 1.002, which is not a highly significant adjustment. This column is the ultimate seasonal description of slip and fall injury frequencies occurring over a typical year. The individual index numbers in this column may be plotted as in

Table 13.5. Results of Dividing Monthly Frequency by Monthly Average

Month	1993	1994	1995	1996	Total of Four Years	Seasonal Index (Average of Four Years)
January	1.482	1.111	1.731	1.207	5.531	1.383
February	1.111	1.296	1.346	.862	4.615	1.154
March	1.111	.926	.962	1.207	4.206	1.052
April	.926	.741	.577	1.035	3.279	.820
May	.741	.556	.769	.690	2.756	.689
June	.556	.556	.385	.690	2.187	.547
July	.741	.556	.577	.517	2.391	.598
August	.741	.741	.962	.862	3.306	.827
September	.926	1.296	.769	1.035	4.026	1.007
October	.926	1.111	1.154	1.207	4.398	1.100
November	1.296	1.482	1.154	1.207	5.139	1.285
December	1.482	1.667	1.539	1.379	6.067	1.517
Totals	12.039	12.039	11.925	11.898	47.901	11.979

Figure 13.5. Inspection of this figure indicates that the frequency of slips is high in the winter, diminishes through the spring, and bottoms out in the summer and then starts to rise again in the fall. Each index number on the graph also indicates the proportion of incidents in a typical year's total that occur in each month. Thus, there are likely to be about twice as many occurrences in January (index 1.38) as in May (index .69). As in the case of trend lines, this technique allows for a certain amount of guarded predictability. Gina can be reasonably certain about events in an upcoming winter, provided that there have been no extreme environmental or organizational changes that would affect the seasonal fluctuations.

13.6 Cycles and Moving Averages. We have already discussed two important aspects of time series: (1) trends, which generally provide an overall description of a large or relatively large number of events, and (2) seasons, which describe a recurring pattern of periodic events. Time series also involve a third concept known as cycles. *Cycles* are oscillating patterns of change that are not necessarily periodic and generally last longer than a year. Cycles are often related to political and economic changes that are external to an organization. Cycles are not as important to the safety and health discipline as they are to

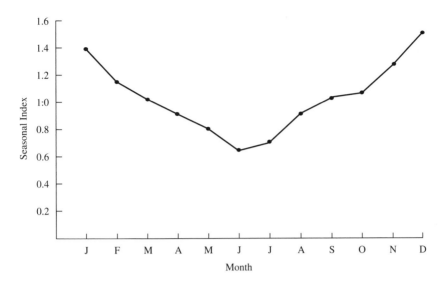

Figure 13.5. Seasonal Pattern of Slips and Falls

economics and business forecasting, but they may be of unexpected value, especially when dealing with public safety.

13.6.1 Procedure. If cyclical patterns exist within a series of events, they may be identified by the *moving average technique*. Moving averages are usually based upon years, but may also be based upon other periods of time. Moving averages are usually based upon groupings of 3, 4, or 5 and are simply an average of the data within the group. A moving average will smooth out the graphic appearance of a series of events and make the presence of cycles more apparent.

13.6.2 Example. Table 13.6 shows vehicular death rates per 100,000 Americans between the years 1943 and 1990. A three-year moving average and a five-year moving average have been applied to the data. Figures 13.6 and 13.7 show the moving averages. The graphs indicate up and down cycles related to vehicular death rates over the years. These cycles are probably explicable in terms of economics, politics, and social change. There are two important concepts to understand about moving averages: (1) when moving averages are used, data are lost at both the beginning and the end of the series; and (2) the larger the individual grouping used (for example, 5 rather than 3), the smoother the cyclical appearance of the graphic will be. Too much smoothing, however, may actually destroy this cyclical appearance. Note that a visually fitted trend line has been drawn onto each cyclical pattern in the figures. This trend line has been determined subjectively, but could have been determined mathematically according to the procedure described in section 13.4.

Table 13.6. Accident Vehicular Death Rates per 100,000 Population

Year	Rate	Three-Year Moving Average	Five-Year Moving Average	Year	Rate	Three-Year Moving Average	Five-Year Moving Average
1943	17.8			1967	26.8	27.1	26.9
1944	18.3	19.1		1968	27.5	27.3	26.8
1945	21.2	21.1	20.8	1969	27.7	27.3	27.0
1946	23.9	22.6	21.7	1970	26.8	27.3	27.0
1947	22.8	22.9	22.3	1971	26.3	26.7	26.8
1948	22.1	22.1	22.6	1972	26.9	26.5	25.6
1949	21.3	22.1	22.7	1973	26.3	25.0	24.5
1950	23.0	22.8	23.0	1974	21.8	23.1	23.6
1951	24.1	23.8	23.3	1975	21.3	21.6	22.7
1952	24.3	24.1	23.5	1976	21.6	21.8	22.2
1953	24.0	23.5	23.6	1977	22.5	21.8	22.6
1954	22.1	23.2	23.5	1978	23.6	23.3	23.0
1955	23.4	23.1	23.2	1979	23.8	23.6	23.0
1956	23.7	23.3	22.6	1980	23.4	23.2	22.6
1957	22.7	22.6	22.5	1981	22.4	21.8	21.7
1958	21.3	21.8	22.1	1982	19.7	20.4	20.8
1959	21.5	21.9	21.5	1983	19.0	19.4	20.0
1960	21.2	21.2	21.4	1984	19.6	19.3	19.5
1961	20.8	21.3	21.7	1985	19.2	19.6	19.5
1962	22.0	22.0	22.4	1986	19.9	19.6	19.7
1963	23.1	23.4	23.3	1987	19.8	19.9	19.6
1964	25.0	24.5	24.5	1988	20.1	19.7	19.5
1965	25.4	25.8	25.5	1989	19.1	19.3	
1966	27.1	26.4	26.4	1990	18.6		

Source: National Safety Council.

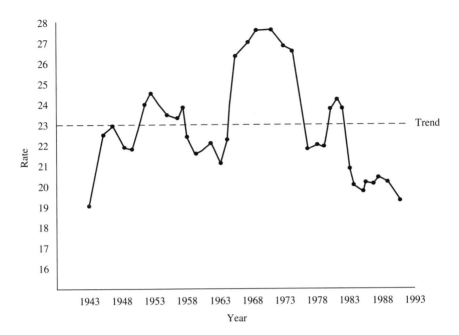

Figure 13.6. Accidental Vehicular Deaths per 100,000 Population: Three-Year Moving Average with Visually Fitted Trend Line

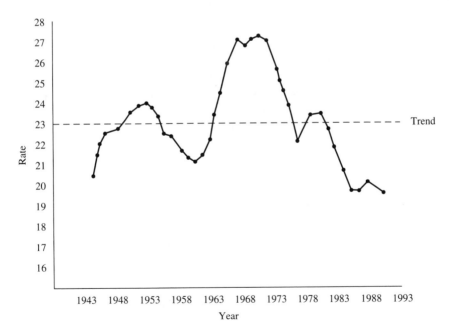

Figure 13.7. Accidental Vehicular Deaths per 100,000 Population: Five-Year Moving Average with Visually Fitted Trend Line

Table 13.7. Ajax Corporation: Accident Frequencies per 1,000 Full-Time Workers

Year	Frequency Rate	Year	Frequency Rate
1975	8.5	1981	6.0
1976	9.1	1982	5.3
1977	9.3	1983	5.2
1978	8.0	1984	5.1
1979	7.9	1985	4.0
1980	7.5	1986	3.8

Problems

13.1. Table 13.7 contains data describing the accident frequency rate of a corporation.

    a. Plot the data upon a graph.
    b. Compute and draw in a trend line for these data.
    c. What are your observations?
    d. Make a prediction.

13.2. Table 13.8 contains data describing the number of safety professionals employed in the Safety Department of the corporation in problem 13.1.

    a. Plot these data on the same graph that you created for problem 13.1.
    b. Compute and draw a trend line for these data.
    c. What are your observations?
    d. Make a prediction.

Table 13.8. Ajax Corporation: Number of Safety Professionals Employed

Year	Number	Year	Number
1975	2	1981	8
1976	2	1982	9
1977	4	1983	9
1978	4	1984	9
1979	5	1985	10
1980	7	1986	10

Table 13.9.   Severity Rates (Rounded) per 200,000 Labor Hours: West Coast Plant

Year	Quarter	Rate	Year	Quarter	Rate
1950	1st	19	1953	1st	27
	2nd	20		2nd	26
	3rd	22		3rd	28
	4th	23		4th	29
1951	1st	23	1954	1st	30
	2nd	23		2nd	30
	3rd	24		3rd	31
	4th	25		4th	32
1952	1st	26	1955	1st	32
	2nd	25		2nd	34
	3rd	27		3rd	35
	4th	28		4th	35

Hint: Use the same y-axis (appropriately labeled) for both accident frequency and number of safety professionals.

13.3. Table 13.9 contains data describing the severity rates of the West Coast Plant of a manufacturing concern.

a. Plot the data on a graph on a quarterly basis.
b. Compute and draw in a trend line.
c. What are your observations?
d. Make a prediction.

13.4. Table 13.10 contains data describing the severity rates of the East Coast Plant of the same corporation as in problem 13.3.

a. Plot these data on the same graph that you created for problem 13.3.
b. Compute and draw in a trend line.
c. What are your observations?
d. Make a prediction.

Table 13.10.   Severity Rates (Rounded) per 200,000 Labor Hours: East Coast Plant

Year	Quarter	Rate	Year	Quarter	Rate
1950	1st	3	1953	1st	4
	2nd	5		2nd	2
	3rd	3		3rd	3
	4th	2		4th	2
1951	1st	4	1954	1st	5
	2nd	5		2nd	3
	3rd	3		3rd	3
	4th	2		4th	3
1952	1st	2	1955	1st	2
	2nd	1		2nd	4
	3rd	5		3rd	3
	4th	3		4th	5

13.5. A corporation safety officer has just organized an Off-the-Job Safety program for employees. He wishes to demonstrate to the Training Department that off-the-job safety training should be emphasized at particular times of the year. Table 13.11 contains monthly data for the last five years describing off-the-job accidents occurring to employees.

    a. Compute and plot the seasonal pattern for these data.
    b. Approximately how many times more accidents are expected to occur in August than in January?

13.6. Mr. Williams, a safety officer for a large manufacturing firm, is interested in finding out whether there is some pattern of accidental injuries occurring over the course of a typical manufacturing day. Mr. Williams already knows that the company works an eight-hour day, five days per week, 50 weeks per year. He has been able to discover that the workers are full-time over the course of a year and that the size of the workforce does not vary greatly. Using stratified sampling, he randomly chooses one Monday, one Tuesday, one Wednesday, one

Table 13.11. Frequency of Off-the-Job Injuries

Month	1991	1992	1993	1994	1995
January	2	3	3	4	2
February	3	4	4	3	3
March	5	5	5	6	4
April	5	5	6	4	5
May	6	6	6	4	5
June	8	8	6	9	8
July	8	8	9	7	7
August	9	8	9	8	8
September	8	6	6	8	7
October	7	6	5	6	7
November	6	4	6	7	7
December	5	4	4	5	6

Thursday, and one Friday from the year that has just passed. He then accesses the accident reports for those days to determine the number of accidents on an hourly basis. The information appears in Table 13.12.

    a. Determine the accident pattern of a typical day.
    b. Which are the most critical times of the day?

13.7. Xenophobia is a mythical nation whose population has remained relatively stable between the years 2000 and 2050. During these 50 years, the country has experienced a number of global wars as well as some severe economic depressions. Leadership of the country is in the form of a military triumvirate, which does not tolerate any "excessive" financial expenditures such as safety and health legislation or even safety and health efforts in manufacturing facilities. The country's labor minister has compiled accident statistics for these 50 years. These data appear in Table 13.13.

    a. Compute a three-year and a five-year moving average.
    b. Plot the moving average that best describes the data and visually fit a trend line.

Table 13.12. Hourly Accident Frequencies

Time	Monday	Tuesday	Wednesday	Thursday	Friday
0800 to 0859	1	2	1	2	2
0900 to 0959	1	2	2	2	1
1000 to 1059	2	3	3	1	1
1100 to 1159	3	3	3	2	3
1300 to 1359	1	2	1	1	1
1400 to 1459	1	2	1	2	2
1500 to 1559	2	3	4	4	3
1600 to 1659	2	4	4	4	4

Table 13.13. Worker Injuries (in Thousands)

Year	Number	Year	Number	Year	Number	Year	Number
2001	3.5	2014	4.1	2027	4.7	2040	3.4
2002	3.3	2015	4.1	2028	4.8	2041	3.5
2003	3.0	2016	3.9	2029	4.9	2042	3.8
2004	3.0	2017	3.7	2030	4.6	2043	4.0
2005	3.1	2018	3.6	2031	4.3	2044	4.4
2006	3.2	2019	3.4	2032	4.1	2045	4.7
2007	3.5	2020	3.5	2033	3.9	2046	4.8
2008	3.7	2021	3.7	2034	3.6	2047	4.9
2009	3.8	2022	3.9	2035	3.5	2048	5.1
2010	4.0	2023	4.0	2036	3.4	2049	5.2
2011	4.4	2024	4.3	2037	3.3	2050	5.2
2012	4.6	2025	4.5	2038	3.2		
2013	4.3	2026	4.6	2039	3.1		

Chapter *14*

Data Collection Procedures

14.1 Discussion. Research requires the collection of information about the physical, social, and managerial environment within an organization. This information, often converted into numbers known as data, must be gathered objectively, thoroughly, methodically, and consistently. This chapter will provide considerations about some of the more prominent data collection techniques used in the behavioral, social, and management sciences.

14.2 Interviews. The average person has been exposed to an interview at some time as either an interviewer or an interviewee. Nevertheless, the average person would be hard put to define exactly what comprises the interviewing procedure. Some essential elements are found in all interviews, however: (1) the interview consists of a series of questions; (2) the interview is on a verbal level; (3) the person(s) who are interviewing have an ascending role, and the person(s) being interviewed have an acquiescing role; and (4) the interview has the mission of gathering information, either specific or general.

14.2.1 Types of Interviews. Interviews follow different procedures and take on a large number of shapes and forms. In all cases, though, the various categories of interviews adhere to the elements just described.

14.2.1.1 Telephone versus Face-to-Face Interview. Interviews are increasingly being performed over the telephone rather than in the more traditional face-to-face manner. Telephone interviews offer the advantages of (1) reaching over long distances, (2) contacting inaccessible people, (3) making use of unusual times of the day, (4) shortening the length of time spent on the interview, and (5) allowing more interviews to be performed in the course of a day.

Face-to-face interviews, on the other hand, offer the advantages of enabling the interviewer to (1) physically assess the interviewee, (2) more readily analyze the psychological characteristics and emotional responses of the interviewee, (3) more readily change the course of the interview in response to subtle changes in the interviewee, and (4) present important physical objects, such as paperwork, tools, or equipment, to be discussed with the interviewee. In the long run, face-to-face interviews are more desirable, especially for matters of great importance.

14.2.1.2 Group and Panel Interviews. A *group interview* is an interview in which more than one interviewee is being asked questions. A *panel interview* is an interview in which more than one interviewer is asking questions. Group interviews can (1) economize on time, (2) build the confidence of the interviewees, and (3) fill in any memory gaps the interviewees may have. Panel interviews offer approximately the same advantages as group interviews as well as the added important advantage of pooling interviewer expertise and experience to produce a more effective interview.

14.2.1.3 Formal versus Informal Interviews. Interviews may be formal with a scheduled appointment set up within the organization's guidelines, or they may occur informally and spontaneously according to an immediate need for information. A formal interview may be a better prepared interview. The informal interview, though not as well prepared, may provide a relaxed atmosphere that will elicit more cooperation and less confused and less anxious responses from interviewees.

14.2.1.4 Structured, Semistructured, and Unstructured Interviews. The trade-off between the rigidity of a structured interview and the flexibility of an unstructured interview will always confront the individual who is responsible for the interview process. A *structured interview* is carefully prepared and researched ahead of time. A list of questions designed by or for the interviewer is usually used. The interviewer usually asks each listed question in turn in a methodical fashion in order to obtain the full range of information that the interview requires. In an *unstructured interview,* the interviewer has only a short list of prepared questions or none at all and relies instead on experience, intuition, and instincts to acquire the needed information. In general, the more complex the issues involved and the more critical the information needed, the more structured the interview should be. Likewise, when rapport and good relationships are of utmost importance, the more unstructured the interview should be. Additionally, an experienced and knowledgeable interviewer may succeed in an unstructured interview whereas a less competent interviewer would do best with a structured interview. A solution to the trade-off of structured versus unstructured is the use of a *semistructured interview.* As a compromise, the semistructured interview uses a list of prepared questions, but the questions are not highly detailed. The list of questions is actually a list of categories or topics that the interviewer covers in an unstructured fashion. In general, unless very specific detailed information is required, the semistructured interview is the most desirable format because it sustains rapport between the parties and still acquires a sufficient amount of facts over the range of information required.

14.2.2 Important Interview Requirements. To conduct a successful interview, the interviewer should adhere to the following guidelines:

1. Prepare adequately for the interview ahead of time. This applies to all interviews whether formal or informal, structured or unstructured.
2. Establish an interview goal and continuously conduct the interview in a manner consistent to attaining the goal. Do nothing that is superfluous to reaching the goal. Do not change the goal during the interview unless some important facts have been exposed that warrant a change.
3. Provide the interviewee with a physical and particularly a psychological environment that will allow for maximum cooperation and accurate responses to questions. Minimal interviewee anxiety is of utmost importance.
4. Keep the interviewee talking, but as much as possible, the talk should be directed toward the interview goals.
5. Speak minimally except to ask questions or make statements that clarify or support the interview process.
6. Ask open-ended questions that require extended verbal responses. Avoid questions that require a simple "yes" or "no" response.
7. Record the information gathered from the interview. Do not rely on memory. Recording is usually in the form of written notes, but may also be in the form of audio or even visual tape recording.
8. Write a report containing the information gathered as well as possible conclusions based upon the information.

14.3 The Observation Technique. Just as interviews are on a verbal level and rely on the auditory channels of a researcher, so observations are on an image level and rely upon the visual channels of the researcher. The *observation technique* is essentially a series of eyesight examinations of a particular environment. Much of what was said about the interview technique also applies to observations, but some additional considerations are offered.

14.3.1 Observing People versus Objects. The observer may look at physical objects in the environment such as machines, tools, or equipment or may observe the activities of people. In most cases, the observer will be concerned with both.

14.3.2 Structured versus Unstructured Observations. As in the case of interviews, observations may be performed either with or without a checklist.

Considering the complexities of a typical environment, it is especially important to use some kind of checklist procedure during an observation. Observations without a checklist may be of some value, however, when looking for new ideas and concepts and theories for future research.

14.3.3 Natural versus Artificial Observations. Observations are usually performed under natural conditions. As an example, an observer visits a particular department in the corporation while the department is performing its usual functions. Observations may also be performed under artificial conditions. As an example, an observer who wishes to observe a particular job function may set up a chair, workbench, and tools in an isolated place and then ask a typical worker to perform his or her usual job.

14.3.4 Participative versus Nonparticipative Observer. The observer may visit a natural environment and unobtrusively observe events, hoping that eventually the members of the environment will become accustomed to her or his presence. This is the usual observational procedure. In some infrequent cases, however, the observer may join the group and observe its members while performing group functions. This procedure has the value of secrecy, putting group members at ease and giving the observer a greater understanding of what functions are about. This participative procedure may prevent the observer from making complete observations, however.

14.3.5 Assigning Observations. The observation technique is often expensive and time-consuming and requires the use of trained, experienced professionals as observers. Because of this, choices about observations must be cost-effective, and observation assignments must be carefully chosen. Observations are assigned based on the following criteria, which are related to sampling procedures:

1. *Time.* Observations may be executed on the basis of time such as once a day, once a week, or every Thursday.
2. *Duration.* Observation may be limited by time duration such as no more than five minutes or one hour.
3. *Time interval.* Observations may be performed at time intervals such as one observation per hour or one observation per five minutes.
4. *Place.* Observations may be assigned according to place such as Department A or the northwest corner of Plant 1.
5. *Function.* Observations may be assigned according to functions and occupations such as assembly operations, manual materials handling, electrician, plumber, or carpenter.
6. *Event.* Observations may be assigned according to specific events such as startup time, quitting time, accidental injury, end of shift or beginning of shift, or overtime.

14.3.6 Measurements. Observations usually end up in some kind of measurement such as time (How long did it take to drive the forklift?), objects and people (How many machines are unguarded? How many workers are not wearing safety glasses?), or actions (How often was the load lifted?). Some observations may end up as opinion ratings (On a scale of 1 to 10, how safety oriented does the department appear?).

14.3.7 Recording. As in the case of interviews, observation results should be recorded in writing and may also be recorded by audio and visual tape.

14.4 Document Searches. In addition to interviews and observations, the safety professional may also gather important information from existing written documents. Documents are usually handwritten or typed notations on paper, but may also take the form of tapes, videos, and computer memory. Document searches are an important part of an activity known as safety auditing.

14.4.1 Varieties. A large variety of documents may be used including but not limited to the following:

accident reports	training records	supervisor's reports
personnel records	blueprints	government reports
medical records	programs	organization charts
inspection reports	OSHA logs	payroll records
budgets	balance sheets	expenditures
profit/loss reports	employee evaluations	supervisor's evaluations
news articles	books	

14.4.2 Advantages and Disadvantages. The advantage of searching through records is that a substantial amount of information can usually be found in one or a few places within a relatively short period of time. In addition, records may be the only source of a particular type of information, especially about past events. The major disadvantage of record searches is the problem of accuracy and veracity. The information and data found in documents were usually collected by someone other than the researcher. Thus, the researcher must trust the competence and truthfulness of the individuals who created the documents. Through physical inspection, the researcher may determine the veracity of some documents such as inspection reports.

14.5 Testing. A test is a determination of the characteristics of animate or inanimate objects. This determination is the result of making a series of observations, usually after manipulating these animate or inanimate objects in some structured, methodical manner. Although the interview and observational techniques might also be considered to be tests, this section will focus only on the types that are familiar to the average person.

14.5.1 Variety of Tests. Tests assume a large variety of forms and shapes. Some test varieties and characteristics are discussed in the following paragraphs.

14.5.1.1 Inanimate Objects or People. To assure safety and health, it is sometimes important to test inanimate objects such as chains, ropes, slings, electrical systems, and sprinkler systems. More frequently, however, people rather than objects will be tested for various reasons related to safety and health.

14.5.1.2 Performance, Oral, Paper and Pencil. Performance tests are administered less frequently than other forms of testing. Nevertheless, some of these tests may be of extreme importance. Some examples of performance tests are forklift operation, crane operation, over-the-road vehicular driving, machine and equipment maintenance, and fire fighting. These are all tests of knowledge combined with sensory motor skills that require a high level of competence, which in turn assures the safety and health of those involved. Oral tests require verbal responses from a testee and may occasionally be required in cases of physical or mental incompetence or illiteracy on the part of the testee. Paper and pencil tests are most frequently used and may require the use of pencil on paper or the use of pencil on machine-graded forms.

14.5.1.3 Objectivity versus Subjectivity. In all cases, a test should be designed so that the tester can arrive at an objectively derived conclusion about the object or person tested. The opportunity to grade objectively is of utmost importance. The best-known example, from the field of education, is the multiple choice test (considered to be objectively graded) as compared to the essay-type test (considered to be subjectively graded).

14.5.1.4 Scope. As mentioned earlier, tests are designed to determine the characteristics of people or things. Usually, these characteristics are grouped into different categories, and the test is designed to deal with one or a few of these categories. Therefore, to be well designed, a test must cover all the characteristics in a category. As an example, a written test related to the category of "safe forklift operator" must cover all characteristics related to safe operation such as maintenance, driving, signaling, and entering trucks.

14.5.1.5 Uniform Testing Procedures. To ensure that all testees are comparable, the test must be administered in the same way to all those who take it. This principle applies to the sequence in which questions are presented and especially to the duration of the test, unless the test has been designed without a time limit. The test environment should also be uniform. Remember also that an inclement, uncomfortable test environment will not allow an accurate testing of the testee's characteristics.

14.5.1.6 Scaling. For comparability among all testees, the test should, if at all possible, be designed so that the final result is a number located on an increasing, equal-interval scale. A traditional percentage score ranging from 0% to 100% would satisfy this requirement.

14.5.1.7 Types of Paper and Pencil Tests. The following are types of paper and pencil tests likely to be used frequently by the safety and health professional:

1. *Intelligence tests.* These tests measure general intellectual ability or a specific intellectual ability such as mathematics.
2. *Aptitude tests.* These instruments measure the testee's potential for learning a particular skill such as tool and die maker.
3. *Achievement tests.* These tests measure the testee's current mental skills in a particular field. Sensory motor skills may be tested as well. Examples include crane operating, traffic laws, and safety rules.
4. *Personality tests.* These instruments measure how the testee may relate to his or her social and managerial environment.
5. *Opinion and attitude surveys.* These surveys measure how the testee feels and what the testee thinks about people, the organization, procedures, and the like.

Several of these tests fall within the jurisdiction of a professional psychologist. They may still be of use to the safety and health professional if a psychologist's aid is enlisted. Others, such as the achievement test and opinion and attitude surveys, may be designed and administered by the safety and health professional.

14.5.1.8 Design of Questions. A well-designed question on a test or survey must have the following characteristics:

1. *Clarity and comprehensibility.* The question must be worded so that it is completely understood.
2. *Pertinence.* The question should be related to the purpose of the test. If a question has nothing to do with the test, it should be eliminated.
3. *Avoidance of leading questions.* The wording of the question should not guide the testee into a response. Example: Why do you like the latest safety posters hung in the plant?
4. *Appropriate format.* Due consideration should be given to various test formats such as completion, true-false, multiple choice, rating scale of 1 to 10, and the like. The best format is one that will best achieve the desired results.
5. *Sensitive areas.* Questions in psychologically and socially sensitive areas may result in evasive or false responses. Examples: What is your income? Do you like your present supervisor? If such questions are imperative, then anonymity may gain the testee's cooperation.

14.6 Safety and Health Applications. Interviews, observations, document searches, and tests may all be used at some time by the safety and health professional for fact finding and data collection. More specifically, an accident investigation, which is a common activity of the safety professional, strongly relies on interviews of the injured person, the supervisor, and witnesses. Gathering accident information via an interview is unavoidable and indispensable because the safety professional is rarely at the scene when an accident occurs. Importantly, but to some lesser extent, the accident investigation relies on the observation of equipment, tools, and the like that are related to the accident. In addition to accident investigations, safety professionals become involved in the Job Safety Analysis (JSA) procedure. This procedure, which analyzes the potential hazards of a particular job, makes extensive use of the observational technique. Sometimes, the safety professional uses the critical incident technique, an interview procedure for asking workers about "near miss" incidents that they have experienced or might conceivably have experienced. Finally, safety professionals may perform safety audits, which are examinations and evaluations of safety programs or parts of safety programs. Safety audits require a substantial amount of document searches plus some interviews and observations.

14.7 Categories. Chapter 2 of this book has already discussed the use of categories. After data collection, categories become indispensable tools for pooling the data into meaningful groups, which in turn provide an understanding of the physical, social, and managerial environment. The definition and delineation of categories should be up to the individual researcher and should vary with the particular research problem at hand. Appendix C provides a Supplementary Record of Occupational Injuries and Illnesses created by the National Safety Council that may give the researcher insights into creating a series of categories of various types.

Chapter 15

System Safety

15.1 Introduction. Safety and health may be divided into two broad categories: *system safety* and, for want of a better term, *traditional safety.* Traditional safety is the discipline that has been addressed throughout this book. Traditional safety is the category most frequently and predominantly used by safety and health professionals. In contrast, system safety is used less frequently and is practiced by a smaller number of safety and health professionals who have some highly specialized interests. Both traditional safety and particularly system safety are involved with research. The differences between their research approaches will be described in the next paragraphs.

15.1.1 Looking Backward or Looking Forward. The most important difference between the two approaches is that traditional safety looks backward in time whereas system safety looks forward. These concepts are sometimes referred to as *ex post facto* (after the fact) and *a priori* (before the fact). Ex post facto traditional safety waits for injuries and diseases to occur. Traditional safety then analyzes the various types of injuries and diseases and their related circumstances and sets up remedial actions to prevent future similar occurrences. This procedure is often criticized for, in effect, "closing the barn door after the horse has run away." Nevertheless, it is the "state of the art" in the safety and health field and in other prominent fields such as medicine and public health. A priori system safety, on the other hand, looks forward and makes conjectures as to what might occur in the future. When the conjectured occurrences, which are usually of an undesirable nature, are analyzed, preventative designs and procedures are implemented to minimize the probability that they will happen.

15.1.2 Areas of Concern. Traditional safety is concerned with large numbers of people who may be workers in various industries or who may be members of the general public. This concern is also extended to property that might be associated with these people. In general, hazard severities exposed to these people and objects vary from low to high but are usually not catastrophic in nature. Usually, loss potential is great but not unbearable. In contrast, system safety is likely to be concerned with smaller numbers of people and with very critical, expensive equipment. Catastrophe potentials are often high and losses may be unbearable.

15.1.3 Extent of Effort and Methods. Traditional safety is often more extensive in its efforts and will seek simple engineering and managerial solutions in response to its concern for the organization it serves. Traditional safety uses well-known parametric and nonparametric statistical tools for its research. System safety, on the other hand, may work more intensively in smaller, more restricted areas and may seek solutions at complex engineering levels. System Safety has many specialized research tools. Only two of the best-known tools will be described in this chapter.

15.2 Risk Assessment and Acceptance Coding (RAAC). The information in this section is a composite derived from Roger L. Brauer's *Safety and Health for Engineers* and Joe Stephenson's *System Safety 2000.** These books in turn have acquired information from the U.S. government's Military Standard 882 B. There are a number of established system safety procedures such as Failure Mode and Effect Analysis (FMEA), Energy Trace and Barrier Analysis (ETBA), and Preliminary Hazard Analysis (PHA). All of these somehow contain the principles related to *Risk Assessment and Acceptance Coding (RAAC).* This section will provide basic information related to these procedures that will aid the safety and health practitioner.

15.2.1 Two Basic Concepts and Definitions. Two definitions must be provided and understood. Stephenson's book states them succinctly as follows: (1) *hazard* is a condition that is a prerequisite to a mishap; (2) *risk* is an expression of the possibility of a mishap in terms of hazard severity and hazard probability. The safety practitioner is concerned not only with the presence of circumstances that can lead to an accident but also with understanding the likelihood that these circumstances will lead to the accident and how severe the consequences of the accident will be. As an example, a hole in the main walkway of a factory is walked over by many employees in the course of a day. The hole is large enough and deep enough to cause an employee to stumble and fall upon the walkway. The hole should be considered a falling hazard. From the

Safety and Health for Engineers by Roger L. Brauer, Van Nostrand Reinhold, 1990. *System Safety 2000* by Joe Stephenson, Van Nostrand Reinhold, 1991.

Table 15.1.   Risk Assessment and Acceptance Coding Matrix

Probability Categories	*Severity Categories*			
	I Catastrophic	II Critical	III Marginal	IV Negligible
A Frequent	IA	IIA	IIIA	IVA
B Probable	IB	IIB	IIIB	IVB
C Occasional	IC	IIC	IIIC	IVC
D Remote	ID	IID	IIID	IVD
E Improbable	IE	IIE	IIIE	IVE

Interpretations:

IA, IB, IC, IIA, IIB, IIIA: Unacceptable situation (needs immediate emergency action).

ID, IIC, IID, IIIB, IIIC: Undesirable situation (quick management decision needed).

IE, IIE, IIID, IIIE, IVA, IVB: Acceptable situation (management review needed).

IVC, IVD, IVE: Acceptable situation (no management review needed).

point of view of risk, the probability of an employee falling is rather high because many employees walk over the hole in the course of a day. Also, from the point of view of risk, the severity of the fall is moderate to high in that sprains or broken bones may result. Suppose now that the hole is in the surface of a catwalk in an upstairs equipment room that is rarely used by employees and is not protected by guardrails. The falling hazard is still the same, but the risk is different. The probability of a fall is substantially lower because employees rarely walk on the catwalk, but the severity of a fall is much higher because an employee who trips because of the hole will go over the edge of the catwalk and land 20 feet below.

15.2.2 Risk Assessment and Acceptance Coding Matrix. RAAC is dependent upon the matrix in Table 15.1. The following definitions apply to terms used in the matrix:

Severity definitions:

  I Catastrophic: May cause death or loss of a facility.
 II Critical: May cause severe injury or illness or major property damage.
III Marginal: May cause minor injury or illness (with lost work day or days) or minor property damage.
IV Negligible: Probably no effect on people's safety and health.

Probability definitions:

A Frequent: Likely to occur frequently.
B Probable: Will occur several times in the life of a facility or system.
C Occasional: Likely to occur sometime in the life of a facility or system.
D Remote: Unlikely but possible to occur sometime in the life of a facility or system.
E Improbable: So unlikely that it can be assumed that occurrence may not be experienced.

To perform this particular procedure in system safety analysis, a safety professional would partially or totally inspect an existing facility, process, or system or review the plans of a proposed facility, process, or system. Using prior experiences and special training, the professional would identify one or more hazards and analyze them in accordance with the probability and severity definitions that are provided with the matrix. The professional then finds the appropriate entry in the matrix in order to decide how acceptable this hazard(s) is to his or her organization. The hazard(s) are then listed in a manner that would be comprehensible to the organization. An example of such a listing appears in the next section.

15.2.3 Example. A manufacturing company has a factory on the East Coast and decides to open a similar factory on the West Coast. The factory has been constructed and machines and equipment installed. The safety officer is asked to inspect the new factory before operations get under way. He discovers the following hazards:

1. Exit doors swing inward rather than outward.
2. Explosion potential in paint locker (no explosion-proof electrical fixtures).
3. Missing guardrail on a three-foot-high stairway.
4. No hot water in employees' restrooms.
5. Punch press unguarded.

Using his judgment, the safety officer rates each hazard as to severity and probability and decides as follows: (1) Exit doors swinging in may lead to the catastrophe of lost lives during a fire. This occurrence may be remote. This hazard is judged to be ID, Undesirable Situation needing quick management decision. (2) The explosion potential in the paint locker may lead to catastrophic loss of life from explosion. This occurrence is considered frequent. This hazard is judged to be IA, Unacceptable Situation needing immediate emergency action. (3) The missing guardrail on the stairs may lead to minor injury. An injurious fall will occasionally happen. This hazard is judged to be IIIC, Undesirable Situation needing quick management decision. (4) No hot water in employees' restrooms creates a remote possibility of mild gastrointestinal illness. This hazard is interpreted as IIID, Acceptable Situation needing management review. (5) The unguarded punch press may lead to an amputation and the occurrence is probable. This hazard is interpreted as IIB, Unacceptable Situation needing immediate emergency action. The safety officer may structure his findings, evaluations, and suggestions as in Table 15.2.

The listing in the table, which will be attached to the safety officer's report, will indicate priorities for management actions. The two identified hazards that call for the quickest action are the paint locker and the punch press. The exit doors, the handrail, and the hot water heater are lesser priorities.

15.2.3.1 Additional Thoughts. It can be seen that this procedure fits well into the scheme of an organization's safety inspection program. Three additional points should be noted about this procedure: (1) it could also be applied to a facility that is already in operation; (2) it would also work on a single process such as a large machine or assembly line; and (3) by reviewing blueprints and other documents, this procedure may be applied at the planning stage of a facility or process. Admittedly, reviewing blueprints will not produce as much information as actually inspecting a finished facility or process, but information may still be acquired that will save on expensive retrofitting during the construction/manufacturing process. Finally, it should be emphasized that this procedure is helpful, but it is also somewhat subjective in nature and relies on a substantial amount of individual judgment.

15.3 Fault Tree Analysis (FTA). The other important system safety methodology is known as *fault tree analysis (FTA).* FTA is a graphic procedure that displays and analyzes the underlying causes or conditions leading to a single

Table 15.2. Hazards in the New Plant

	Hazard	Consequences	Severity	Probability	Suggestion
1	Exit doors swing in	Employees cannot exit, die in fire	Catastrophic	Remote	Change direction of door swing
2	Explosive atmosphere in paint locker	Explosion kills workers	Catastrophic	Frequent	Install explosion-proof fixtures
3	Missing handrail on low stair	Fall from low elevation, injury	Minor	Occasional	Install handrail
4	No hot water in restrooms	Employee intestinal illness	Marginal	Remote	Repair hot water heater
5	Unguarded punch press	Amputation	Critical	Probable	Install guard

unwanted event. The information that follows in this section was acquired from the same books mentioned in the prior section. The information presented here is not as detailed as a systems safety engineer would require, but should be adequate for the needs of a typical safety and health professional.

15.3.1 Construction and General Procedure. An FTA displays a number of related events or conditions in a branching chainlike pattern. The events start at the bottom of a page and work their way upward to produce an explanation for a single, major, undesirable event at the top of the page. The tree is analogous to several lines of standing dominoes converging at a single point with only one domino. If the end domino of some or all of the lines of dominoes is pushed over, the chain reactions will eventually knock over the single domino at the point. An example of an FTA construction will be provided, but first some FTA symbols must be explained.

15.3.2 Symbols and Concepts. Figure 15.1 shows and explains the various symbols that are used:

a. A rectangle symbolizes an event, usually of an undesirable nature. An undesirable event may be classified as a fault event in which something did not perform correctly or as a failure event in which something did not perform at all. These events need to be further described by other events.
b. A circle is also an event, but it requires no further description by other events.
c. A diamond is an undeveloped event. The researcher does not wish to provide any further descriptive events.
d. A house is an event that is normal and is usually found as part of the system being studied.
e. An oval is sometimes placed next to an event and connected to it with a line. The oval may contain additional information about the event.
f. A triangle is a transfer device. The FTA sometimes becomes large and complex. The triangle directs attention to another sheet of paper where the tree continues.
g. An OR gate receives two or more inputs from events below and produces a single output to an event above. The occurrence of one or more inputs into an OR gate can cause the event above.
h. An AND gate receives two or more inputs from events below and produces a single output to an event above. All of the inputs going into the gate must occur in order to cause the event above.

15.3.3 Example. Figure 15.2 is an FTA provided during a 1989 training session conducted by the OSHA Training Institute. By following the step-by-step construction of the FTA, we will learn how to design our own FTA when solving safety problems.

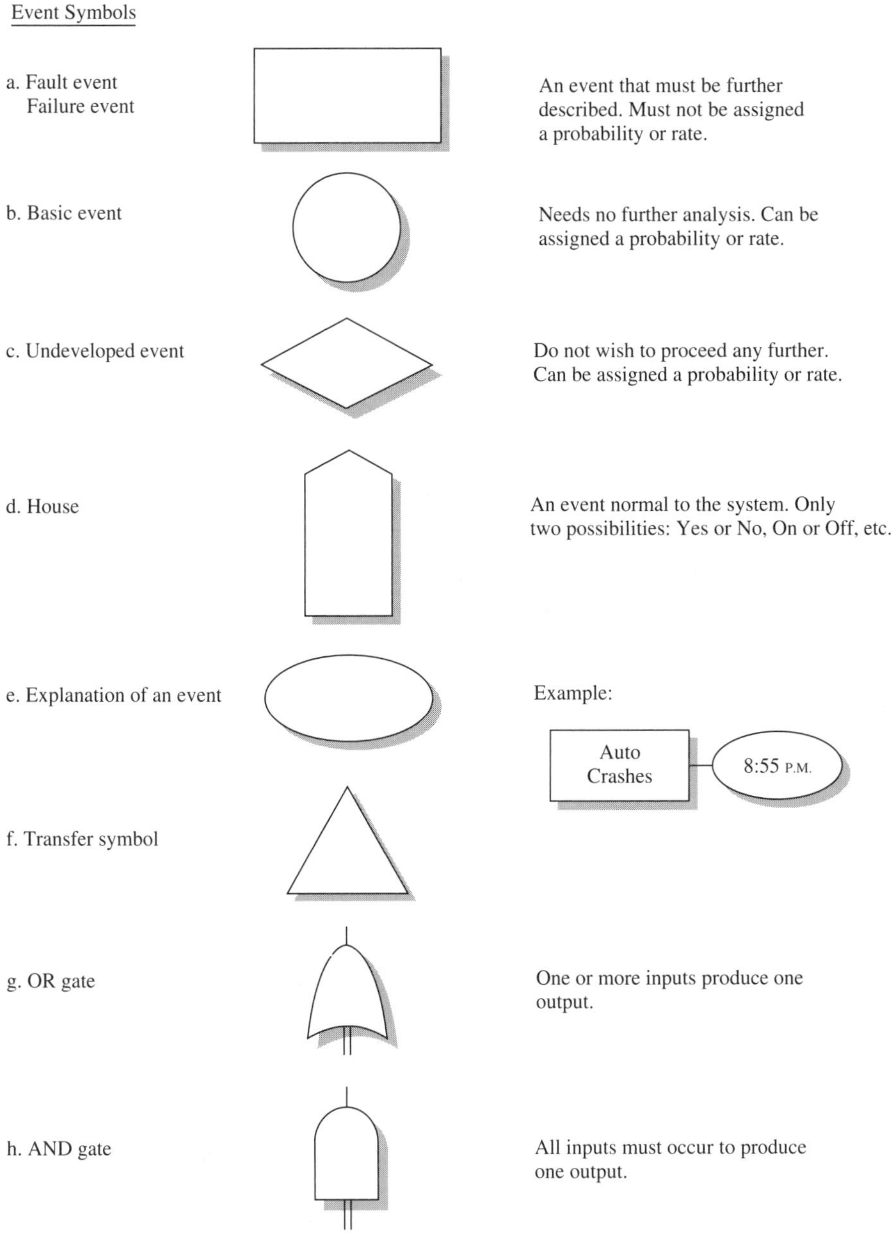

Event Symbols

a. Fault event   An event that must be further
   Failure event   described. Must not be assigned
                                                        a probability or rate.

b. Basic event   Needs no further analysis. Can be
                                                        assigned a probability or rate.

c. Undeveloped event   Do not wish to proceed any further.
                                                        Can be assigned a probability or rate.

d. House   An event normal to the system. Only
                                                        two possibilities: Yes or No, On or Off, etc.

e. Explanation of an event   Example:

                                                        | Auto Crashes | 8:55 P.M. |

f. Transfer symbol

g. OR gate   One or more inputs produce one
                                                        output.

h. AND gate   All inputs must occur to produce
                                                        one output.

Figure 15.1.   Fault Tree Analysis

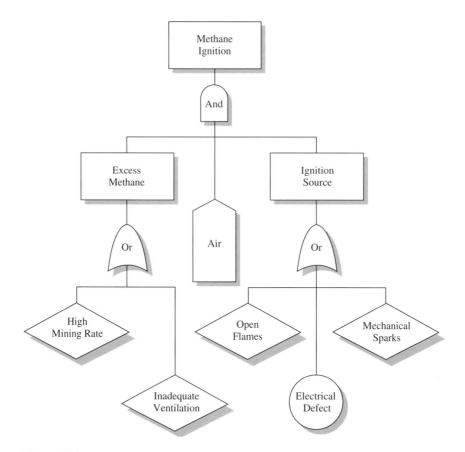

Figure 15.2.   Underground Coal Mine Explosion

1. Construction begins at the top. Although causation flows upward from the bottom of the page, the FTA is constructed from the top down and begins with a single event. Because this technique is relatively expensive and time-consuming, only a critical or catastrophic event should be used. In this case, the safety officer of a mine is concerned about a catastrophic underground explosion that may kill many workers at once. The safety officer knows that underground mine explosions are caused by the ignition of methane gas; thus, "Methane Ignition" is the first occurrence at the top and is placed in a rectangle to signify that this event needs further analysis.
2. Beneath the top event, an AND or OR gate must be placed. In this case the safety officer chooses an AND gate based upon his knowledge of fire chemistry. He knows that a fire or explosion can only occur if three

prerequisite conditions exist: adequate fuel, adequate oxygen, and a source of ignition. He therefore places the AND gate below the top event, indicating that all of the events below must occur in order to have an explosion.

3. The events at the next level—Excess Methane, Air, and Ignition Source—are treated differently. Excess Methane and Ignition Source are placed in rectangles because their occurrence needs further explanation. Air, however, is natural and normal to the mine environment. It is taken for granted and thus is placed in a house. No further explanation of the presence of Air is necessary, but both Excess Methane and Ignition Source require further explanation.

4. At the next lower level, OR gates are used beneath Excess Methane and Ignition Source. These gates indicate that the events below them are independent of each other and do not have to occur simultaneously to produce the events above.

5. Beneath the Excess Methane OR gate, the safety officer places two events or conditions, High Mining Rate and Inadequate Ventilation, which he knows can lead to Excess Methane. As noted, these events are independent of each other; that is, either one can cause a buildup of methane. These events are placed within diamonds because the safety officer does not wish to analyze them further.

6. Beneath the Ignition Source OR gate, the safety officer places three independent events, each of which is known to have the capability of igniting methane. Open Flames and Mechanical Sparks are placed within diamonds because the safety officer does not wish to analyze them any further. Electrical Defect is placed in a circle because the safety officer believes there is no need to analyze it further.

15.3.4 Interpretation. After constructing and examining the tree, the researcher will detect some of the root causes of an explosion and will understand that by controlling mining rate, ventilation, and flame and spark generation, the mine can guard against explosions.

15.3.5 Discussion. This method is an effective procedure for visualizing causations, but as in the case of risk analysis, a considerable amount of subjectivity is introduced into the methodology. Note that although an undesired event is usually placed at the top, a desirable event or circumstance may be used instead. In the case of the FTA in Figure 15.2, the top event could be Mine Atmosphere Is Explosion-Free. Appropriate events and gates would then be used beneath this positive top event. Finally, there are provisions for quantifying and for determining probabilities of various events within the tree. Based upon prior research and experience or based upon judgment, probabilities for various events may be inserted into the tree. A series of additions or multiplications will provide

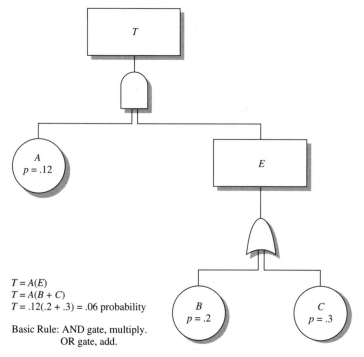

$T = A(E)$
$T = A(B + C)$
$T = .12(.2 + .3) = .06$ probability

Basic Rule: AND gate, multiply.
OR gate, add.

Sometimes OR gates will add up to probabilities of more than 1 (which is not possible). This means events are not mutually exclusive. Solution: Subtract from 1, then multiply.
Example: $p = .4 + p = .3 + p + .5 = p = 1.2$
Change to $p = 1 - [1 - .4(1 - .3)1 - .5] = .79$

Figure 15.3   Fault Tree Quantification

the probability of occurrence of the top event. Figure 15.3 explains the computational procedure. One of the most important rules is that AND gates require multiplication and OR gates require addition.

Problems

15.1 A recent safety inspection in a plant reveals the following write-ups:

    a. Electrical receptacles in the West Wing are ungrounded.
    b. Compressed gas cylinders are not secured from falling over.
    c. Exit doors in a plant are not marked with exit signs.

d. The ventilating fan of the ventilator above the paint dip operation is not operating.
e. The sprinkler system has not received its monthly maintenance inspection.

Perform a risk analysis and acceptance procedure upon these write-ups.

15.2 A manufacturing company has a large number of almost identical punch presses that are capable of severing an operator's hand. The presses are equipped with state-of-the-art protection in the form of point of operation guards and antirepeat mechanisms. The safety officer still wishes to analyze these machines even though they have protection.

a. Construct a fault tree analysis on a typical machine. The top event should be labeled Hand Crushed by Descending Die.
b. Perform a risk analysis and acceptance based upon the failure of the guard and upon the failure of the antirepeat mechanism.

15.3 An aircraft is powered by a traditional gasoline engine.

a. Perform research about the basic operation of a gasoline engine. Construct an FTA based upon the major systems of a typical engine. The top event should be Engine Fails during Flight.
b. Perform a risk analysis and acceptance procedure upon the engine.

15.4 A worker is required to use a solvent to clean the inside of a rusted steel tank that has only one small entry port. The worker is observed unconscious at the bottom of the tank. The worker was wearing a safety harness attached to a line leading outside the tank. No coworker was found anywhere around the outside of the tank. Construct an FTA.

Chapter *16*

Use of Computers

■ BY ROGER O. RAMSAY, MS MA

16.1 Introduction. Electronic calculators perform only the calculations specified by the researcher. In addition, when complex statistical analysis is performed, hand calculation can become very tedious and time-consuming. The statistical analysis can be made simpler and conducted faster through the use of computers. In this chapter we describe the general principles behind the use of computers in conducting statistical analysis and the general features of statistical software programs that aid in data analysis, statistical problem solving, and hypothesis testing.

16.2 The Role of the Computer in Statistical Analysis. Many handheld calculators can perform only one task at a time and perform that task in only one way. Other hand calculators are programmable and are capable of displaying numbers, letters, and some special symbols. In addition to a diversity of scientific functions, many hand calculators have a programming function for repeat and complex calculations. The use of algebraic logic in writing programs to perform repeat and/or complex calculations can become very time-consuming and limited, however.

A computer is much more versatile. It can perform statistical analysis, help conduct economic forecasting, and do desktop publishing simultaneously in a multitasking environment. The computer can perform multiple tasks because its actions can be controlled by programs that give different sets of instructions that are executed at specific times.

With a hand calculator, we can hand-tabulate and calculate interview responses and frequencies, but the process is time-consuming and inefficient. In comparison, a modern microcomputer with an Intel Pentium CPU (central processing unit) can perform a sophisticated statistical analysis on a vast volume of data in an infinitesimal amount of time. The importance of microcomputers in research is not limited to qualitative or quantitative data analysis. Microcomputers also play a vital role at various stages of the research process: they are used for telephone interviewing, sampling, coding for content analysis, cleaning and printing a data file, conducting the statistical analysis, hypothesis testing, and desktop publishing the final document.

As microcomputers have become less expensive, more accessible, more interactive, and more powerful with more memory capacity and larger storage space, the microcomputer has replaced the mainframe and the minicomputer system as the central research tool. In addition, microcomputers offer the following advantages for researchers: (1) battery-powered microcomputers can be taken directly to the field, where they can be used to enter data and notes directly and to access current safety regulations; (2) a wide variety of useful software applications packages are available in the microcomputer format; and (3) owing to the low cost of microcomputers, researchers can afford to purchase them outright and avoid the stiff timesharing costs usually associated with mainframe and minicomputer systems. Therefore, our discussion will emphasize the use of microcomputer statistical programs instead of the traditional mainframe or minicomputer system programs.

Learning to use the microcomputer for statistical analysis is now easier than ever before. In a short period of time, the user can access the almost unlimited statistical power and presentation-style graphics of these easy-to-use statistical programs. Without much prior computer experience and with the benefit of on-line programs HELPS, a researcher can quickly and easily learn to use these software programs to perform complex statistical data analysis.

16.3 The Computer Software Program. Software packages of prewritten computer programs are available for conducting practically any statistical analysis. They are widely used and commercially available and are the software programs used in universities and colleges, in scientific research laboratories, in medical research facilities, in governmental institutions, and in business and industrial organizations.

All software programs are written in one of several computer languages that the microcomputer can "understand." Overall, computer languages have a small "vocabulary" and a simple, unambiguous syntax (a set of rules for constructing statements in the language). To write a productive computer program, it is necessary to learn the syntax of the language and apply it to the construction of statements using the language's vocabulary. Fortunately, we do not need to learn how to write a computer program in order to make effective use of a computer. Once a program has been written to instruct the computer to do a task (such as to compute the mean and variance of a set of numbers), anyone can use that program to perform the task without having to understand exactly how the program was written or what the individual statements in the program mean. Nevertheless, it is important to be firmly grounded in the theoretical concepts underlying statistical analysis and to be able to recognize the appropriate application and possible limitations of each statistical test employed.

16.4 Introduction to Statistical Software Programs. A microcomputer statistical package is a set of analytical programs, written for the common purpose of statistical analysis and constructed to be easy for nonprogrammers to use. The increasing use of microcomputers has led to greater demand for various types of software to assist in the management and analysis of information in all fields. Statistical packages allow relatively unskilled users to benefit from utilizing computers in data analysis and statistical problem solving. The standardized nature of a statistical package ensures that the results of the computerized analysis will be reliable and that the time spent implementing the statistical programs will be minimized.

While many statistical packages have been used for years on mainframe and minicomputer systems, over the past decade numerous packages have been developed for use with microcomputers. Many of these statistical packages were designed for specific research interests such as medicine and the health sciences, mathematical probability, reliability and quality-control engineering, social sciences and educational research, and economic/financial forecasting. Others were designed to accommodate broader or more general aspects of data analysis and statistical problem solving. Some of these more broad-based statistical packages that have been written for microcomputers include SPSS/PC=, Minitab, SAS, BMDPC, and SYSTAT.

All of these statistical packages use similar methods to give instructions about the analysis to be performed. In a command-driven package, the user constructs statements to indicate what action should be taken. In a menu-driven package, the user is presented with a list of actions from which to choose. While a command-driven package requires a knowledge of its particular syntax in order to run a program, a menu-driven package requires only a knowledge of what keystrokes to make (i.e., "buttons to push") in order to run the program. Therefore, menu-driven packages are generally much easier to use. Nevertheless, using either type of software package is far more efficient than writing the entire computer program itself.

Whichever microcomputer statistical package is selected, the data can be processed in an interactive mode, where the user can give instructions one at a time, see the results, and then give further instructions to the microcomputer. In this way, the user can decide what the next instruction should be on the basis of the results already seen. All of these current statistical programs are designed to analyze and summarize empirical data using common statistical analysis techniques. They can be used for such tasks as performing a factor analysis on a series of variables, testing a hypothesis or a path model, or using analysis of variance to test the significance in an experiment measuring treatment effects.

An example of a modern statistical program is Minitab Release 10 for Microsoft Windows. It is a powerful statistical package that provides a wide range of basic and advanced data analysis capabilities that are seen in many statistical software programs today. With Minitab's well-designed user interface, this program is accessible to both university students and faculty with a wide variety of academic backgrounds and experience. At a glance, Minitab Release 10 for Microsoft Windows offers the following:

- Comprehensive statistics capabilities, including exploratory data analysis, basic statistics, regression, analysis of variance, multivariate analysis, non-parametrics, time series, cross-tabulations, simulations, and distributions.
- High-resolution graphics that enable the user to produce a comprehensive array of graphs, from simple data analysis graphs to impressive, presentation-quality graphs, including three-dimensional graphs. The user can even edit the graphs to customize them and identify points on scatter plots by clicking on them.
- Quality-control and improvement features including Pareto charts, statistical process control charts, process capability graphs, probability plots, analysis of means, cause-and-effect (fishbone) diagrams, design of experiments capabilities to generate and analyze full and fractional factorial designs, Plackett-Burman designs, and response surface designs.
- Powerful data management capabilities including importing data from other versions of Minitab, spreadsheets, databases, and test files, or pasting it from the Clipboard. Data can be linked to another application or another part of the Minitab worksheet using Dynamic Data Exchange (DDE).
- A macro facility that allows the user to write a program of Minitab commands with DO-loops and IF statements to automate repetitive tasks or extend Minitab's functionality. The user can write macros that execute just like a Minitab command, with arguments and optional subcommands.
- A Windows graphical interface that provides an easy-to-use and efficient work environment.

Future statistical programs are anticipated to have these additional desirable features: (1) continued progress toward comprehensive coverage of statistical analysis techniques as well as integration with other programs like database management programs and desktop publishing programs; (2) the use of an interactive environment that allows the user to rotate the results of an analysis visually; (3) further improvements in graphics to take advantage of color, bit-mapped graphics, and three-dimensional displays; (4) user-friendly interfaces that assist in specifying the analysis to be performed and in identifying and correcting misspelled words and misplaced delimiters; and (5) artificial intelligence capabilities that can help in choosing the appropriate statistical test and assist in interpreting the results of an analysis.

Appendix A

Mathematics Calculations

The following mathematical procedures should be more than sufficient for understanding the examples in the text and for solving the problems at the end of the chapters.

Signs

+	plus	$>$	greater than $(8 > 7)$
−	minus	$<$	less than $(7 < 8)$
×	multiply	$\geq$	equals or greater than
÷	divide	$\leq$	equals or less than
=	equals	||	absolute value (ignore the sign) $\lvert+2\rvert = 2, \lvert-2\rvert = 2, \lvert0\rvert = 0$
$\approx$	approximately equals		
$\cong$	approximately equals		
!	factorial (multiply by next lower numbers) $4! = 4 \times 3 \times 2 \times 1$ Note: $0! = 1$		

Addition

Addition Rules: Add similar signs; subtract dissimilar signs. The sign of the larger number predominates.

$1 + 2 = 3$	$(-1) + (-2) = -3$	$0 + 2 = 2$
$(+1) + (+2) = +3$	$(-1) + (+2) = +1$	$2 + 0 = 2$

148

Subtraction

Subtraction Rules: Change the second and third signs; then follow rules of addition.

5 – 4 = 1   (–2) – (+4) = –6   0 – 2 = –2

(+5) – (–4) = 9   (–2) – (–7) = 5   2 – 0 = 2

(+5) – (+4) = 1

Multiplication

Multiplication Rules: Similar signs produce a plus; dissimilar signs produce a minus.

7 × 5 = 35   (7) · (5) = 35   (+7)(–5) = –35

7 · 5 = 35   (+7)(+5) = 35   (7)(0) = 0

7(5) = 35   (–7)(+5) = –35   (0)(7) = 0

(7)(5) = 35   (–7)(–5) = 35

Division

Division Rules: Similar signs produce a plus; dissimilar signs produce a minus.

4 ÷ 2 = 2 $\dfrac{+4}{+2} = 2$ $\dfrac{+4}{-2} = -2$

4/2 = 2 $\dfrac{4}{2} = 2$ $\dfrac{-4}{+2} = -2$

$2\overline{)4} = 2$ $\dfrac{0}{2} = 0$ $\dfrac{-4}{-2} = 2$

$$\frac{5!}{3!} = \frac{5 \times 4 \times 3 \times 2 \times 1}{3 \times 2 \times 1} = 5 \times 4$$

Order of Operations

Rule: Operations within [ ] and ( ) are performed first.

$3(2 + 1) = 3(3) = 9$

$$\frac{8(2)}{4} = \frac{16}{4} = 4$$

$3 + 3(2 + 1) = 3 + 3(3) = 3 + 9 = 12$

$$\frac{4+8(2)}{4} = \frac{4+16}{4} = 5$$

$2[(8 \times 2)(9 + 1)] = 2[(16)(10)] = 2(160) = 320$

Exponents

Rules:

Multiplying: Add exponents.   Minus: Change to reciprocal; place under 1: $\underline{1}$.

Dividing: Subtract exponents.   Fraction: Change to radical $\sqrt{\ }$.

$2^0 = 1$

$$\frac{2^3}{2^2} = 2^1 = 2$$

$2^{1/3} = \sqrt[3]{2}$

$2^1 = 2$

$-2^2 = -2(-2) = 4$

$2^{2/3} = \sqrt[3]{2^2}$

$2^2 = 2 \times 2 = 4$

$-2^3 = -2(-2)-2 = -8$

$2^{-2} = \dfrac{1}{2^2}$

$2^3 = 2 \times 2 \times 2 = 8$

$2^{1/2} = \sqrt[2]{2} = \sqrt{2}$

$2^{-3} = \dfrac{1}{2^3}$

$(2^3)(2^2) = 2^5$

$2^{.5} = 2^{1/2} = \sqrt{2}$

$2^{-1/2} = \dfrac{1}{\sqrt{2}} = \dfrac{1}{2^{1/2}}$

Radicals

$\sqrt[2]{4} = \sqrt{4} = 2$ $\sqrt[3]{27} = 3$

To find the square root of large numbers, proceed as follows:

Find $\sqrt{8,000,000}$. Move 4 decimals over (always in 2s), making $\sqrt{800} = 28.284$.

Since we moved the decimal 4 places to the left, we now must move the decimal 4/2 places to the right:

$$\sqrt{8,000,000} = 2,828.4$$

To find $\sqrt{76,900,000}$, move 6 decimals over:

$$\sqrt{76.9} \quad \text{(find)}$$

$$\sqrt{77} = 8.7750$$

$$\sqrt{76} = 8.7178$$

$$\text{Difference} = \ .0572$$

$$\frac{\times \quad .9}{.05148}$$

$$\text{Add: } \sqrt{76} = \frac{8.7178}{8.76928}$$

Move the decimal 3 places to the right: $\sqrt{76,900,000} = 8,769.28$.

Handling Zeros

$1,000,000 = 1(10^6) = 10^6 = E6$ $\qquad \dfrac{1}{1,000,000} = .000001 = 1(10^{-6}) = 1E - 6$

$2,000,000 = 2(10^6) = 2E6$ $\qquad \dfrac{4}{1,000,000} = .000004 = 4(10^{-6}) = 4E - 6$

$2,500,000 = 2.5(10^6) = 2.5E6$

$100,000 = 1(10)^5 = E5$

$2(10^6) + 4(10^6) = 6(10^6)$ $\qquad\qquad 2(10^{-6}) + 4(10^{-6}) = 6(10^{-6})$

2E6 + 4E6 = 6E6

$4(10^6) - 2(10^6) = 2(10^6)$

4E6 − 2E6 = 2E6

$4(10^6) \div 2(10^6) = 2$

4E6 ÷ 2E6 = 2

$4(10^6) \times 2(10^6) = 8(10^{12})$

4E6 × 2E6 = 8E12

2E − 6 + 4E − 6 = 6E − 6

$4(10^{-6}) - 2(10^{-6}) = 2(10^{-6})$

4E − 6 − 2E − 6 = 2E − 6

$4(10^{-6}) \div 2(10^{-6}) = \dfrac{1}{2}$

$4E - 6 \div 2E - 6 = \dfrac{1}{2}$

$4(10^{-6}) \times 2(10^{-6}) = 8(10^{-12})$

4E − 6 × 2E − 6 = 8E − 12

Appendix **B**

Chi Square (χ^2) Technique

B.1 Introduction. Since a large number of procedures in this book rely on the chi square (χ^2) technique, this appendix is written as a supplement to and enlargement of the χ^2 examples in the text.

B.2 Categories and Frequencies. χ^2 is dependent upon categories rather than continuous measurement scales. When these categories are used, they are described by frequencies of occurrence. As an example, in a group of 100 people, 75 are men and 25 are women. For the purposes of χ^2, the frequencies of occurrence would be placed in a table as follows:

	Men	Women	Total
Frequency	75	25	100

Note that their total, 100, is also placed in the table as an essential part of χ^2. It should also be noted that numbers on a scale may be changed into frequencies and used in χ^2. As an example, a test whose scores may range from 0 to 100 is administered to 100 people. The range of scores of the 100 people may be converted to frequencies such as the following:

	Score of 50 or More	Score of 49 or Less	Total
Frequency	60	40	100

B.3 Observed versus Expected. An additional important consideration related to χ^2 is the concept of normal expectancy. This technique considers not only the frequencies occurring within a particular category (the *observed*), but also the frequencies that should normally occur within the category (the *expected*). Thus, χ^2 is referred to as a study of the observed versus the expected. Like other inferential statistical methods, χ^2 serves the ultimate purpose of discovering whether the results of a study are due to chance occurrence or are the result of a true difference between groups.

B.4 Types of χ^2. There are two major categories of χ^2. The first is used when a comparison is made within a single group. The second is used when comparing two or more groups.

B.4.1 Comparison within a Single Group. Comparisons within a single group usually rely upon a theoretical expected distribution, usually stated in round numbers, compared to an actual distribution observed by the researcher. As an example, in a company workforce, which is divided approximately equally between men and women, 102 workers report low back pain. Of these 102 workers, 66 are men and 36 are women. A researcher wants to find out if this is a usual occurrence. He sets up the following table.

| | Workers With Low Back Pain | | |
	Men	Women	Total
Observed	66	36	102
Expected (50:50)	51	51	102

As this table indicates, the researcher forms an assumption that a 50:50 distribution of males and females reporting low back pain is expected to occur. A test is performed to see if this assumption is correct. In the χ^2 procedure, expected frequencies are subtracted from observed frequencies, and the result of the subtraction is then squared. The squared result is then divided by the expected frequency, and a grand total is taken of the results of these operations. This procedure is shown in Table B.1.

The final result is a χ^2 of 8.82. The next step is to refer this number to a χ^2 table (Table D.4 in Appendix D) to determine if the 50:50 assumption is correct. In order to use this table, the degrees of freedom of this study must first be determined. The general rule for degrees of freedom (*df*) is to take the number of columns (*k*), subtract 1, and multiply by the number of rows (*r*) minus 1. Thus $df = (k - 1)(r - 1)$. The *df*, which is a statement about the number of deviations between O and E that are free to vary, must be determined for all χ^2 calculations.

Table B.1. Workers with Low Back Pain

	Men	Women	Total
O	66	36	102
E (50:50)	51	51	102
$O - E$	+15	−15	
$(O - E)^2$	225	225	
$\dfrac{(O - E)^2}{E}$	4.41	4.41	

$$\text{Basic formula: } \chi^2 = \Sigma \frac{(O - E)^2}{E}$$

$$\chi^2 = 4.41 + 4.41 = 8.82$$

$$\text{Degrees of freedom } (df) = (k - 1)(r - 1) = (2 - 1)(2 - 1) = 1\ df$$

Note: χ^2 is the symbol for *chi square*; Σ is the symbol for "sum of."

For this particular situation, $df = (2 - 1)(2 - 1) = 1\ df$. The value of 1 df in Table D.4 is either 3.84 at the .05 (5%) level of significance or 6.64 at the .01 (1%) level of significance. Generally, the 5% level of significance should be acceptable for safety research. Since the determined value of 8.82 for χ^2 is greater than the table value (3.84) for 1 df at the 5% level of significance, the assumption that there is an even 50:50 distribution between males and females is not correct. It is concluded with at least 95% confidence that males are more likely to experience low back pain than females.

B4.2 Further Discussion. The prior study is considered a 2 × 2 (two columns and two rows) design. Other designs based upon probability distributions may be 3 × 2 (three columns and two rows), 4 × 2 (four columns and two rows), and so on.

As another example, after reading about a recent catastrophic airline crash, 100 citizens are asked to choose the mode of travel they would prefer for a long trip. The results are shown in Table B.2. As the table indicates, the researcher had formed an assumption that the citizens' preferences would be evenly distributed over the four modes of travel (25:25:25:25). The χ^2 result is 15.92. The df of this study is $(k - 1)(r - 1) = (4 - 1)(2 - 1) = 3\ df$. Table D.4 indicates a value of 7.82 for 3 df at the .05 level of significance. Since the resulting χ^2 value of 15.92 exceeds 7.82, we state with at least 95% confidence that the citizens' preferences are not evenly distributed and that the citizens are wary of airline travel.

Table B.2. Preferred Travel Mode

	Personal Auto	**Railroad**	**Bus**	**Airline**	**Total**
O	33	31	28	8	100
E	25	25	25	25	100
(25:25:25:25)					
$O - E$	8	6	3	−17	
$(O - E)^2$	64	36	9	289	
$\dfrac{(O - E)^2}{E}$	2.56	1.44	.36	11.56	

$$\chi^2 = \Sigma \frac{(O - E)^2}{E} = 2.56 + 1.44 + .36 + 11.56 = 15.92$$

$$df = (4 - 1)(2 - 1) = 3df$$

Table B.3. Workers with Low Back Pain

	Men	**Women**	**Total**
O	66	36	102
E (66⅔:33⅓)	68	34	102
$O - E$	−2	2	
$(O - E)^2$	4	4	
$\dfrac{(O - E)^2}{E}$	.06	.12	

$$\chi^2 = .06 + .12 = .18$$

$$df = (2 - 1)(2 - 1) = 1\ df$$

B4.3 Final Example. In prior discussions, the probable distributions were all equal, such as 50:50 or 25:25:25:25. It is possible to have unequal distributions such as 75:25 or 2/3:1/3 and the like. In the case of men versus women and low back pain, a χ^2 based upon 66⅔:33⅓ might be applied. Table B.3 presents the calculation.

The assumption is that low back pain is distributed as 66⅔ male and 33⅓ female. Since the resulting χ^2 of .18 does not exceed the table value of 3.84 at the 5% level, it is concluded that the distribution is truly $66\frac{2}{3}:33\frac{1}{3}$ and that men experience low back pain twice as often as women.

B.5 Comparison between Groups. The second χ^2 procedure studies characteristics between two or more groups, rather than characteristics within a group.

B.5.1 Two Groups and Two Characteristics. Generally, a researcher will be making comparisons of two groups and two characteristics. This is the technique used most frequently at the beginning of this book.

As an example, a safety officer has analyzed 200 reportable injuries occurring to 200 workers in her corporation over the last year. She was able to determine the age of the workers and whether the injury incurred caused a loss of work time or no loss of work time. She is interested in the effects of maturity upon accident experience. She sets up the information as follows:

	Injured Workers/Number of Accidents		
	Age 30 or Lower	Age 31 or More	Totals
Lost-time accidents	40	50	90
No lost-time accidents	90	20	110
Totals	130	70	200

In order to perform a comparison between the two age groups, the safety officer states the assumption that there is no difference between the safety performances of the younger group and the older group. A χ^2 using a specially designed formula appears in Table B.4.

The final χ^2 result is 28.77. This number exceeds the table value of 3.84 at 1 df at the .05 level. Therefore, the original assumption that there is no difference between the age groups is not correct, and the researcher concludes that the groups are indeed different and that age has an influence upon the frequency as well as the severity of accidents.

B.5.2 More Than Two Groups and Characteristics. Less often, large numbers of units are studied and compared. As an example of a 3 × 3 study, consider the following. A corporation has three departments of approximately equal size that do extensive grinding. Workers in these departments use personal protective equipment to control eye injuries from flying objects. Eye injury accidents for the three departments as related to the type of eye protection being used are recorded in Table B.5, which also includes a χ^2 analysis.

Table B.4. Injured Workers/Number of Accidents

	Age 30 or Lower	Age 31 or More	Totals
Lost time	40(A)	50(B)	90(A + B)
No lost time	90(C)	20(D)	110(C + D)
	130(A + C)	70(B + D)	200 = N

$AD = 20(40) = 800$

$BC = 50(90) = 4,500$

N = total of numbers within cells = $40 + 50 + 90 + 20 = 200$

$$\chi^2 = \frac{N\left(|AD - BC| - \frac{N}{2}\right)^2}{(A + B)(C + D)(A + C)(B + D)}$$

$$\chi^2 = \frac{200\left(|800 - 4,500| - \frac{200}{2}\right)^2}{(90)(110)(130)(70)}$$

$$\chi^2 = \frac{200\left(3,700 - \frac{200}{2}\right)^2}{90,090,000}$$

$$\chi^2 = \frac{2,592,000,000}{90,090,000}$$

$$\chi^2 = 28.77$$

$$df = (k - 1)(r - 1) = (2 - 1)(2 - 1) = 1 \ df$$

Note: $|AD - BC|$ means "absolute"; ignore the resulting sign.

As in all other χ^2 techniques, this design requires expected frequencies as well as observed frequencies. A special procedure is used to determine the expected frequencies. To determine E, the product of the column and row totals of the O frequency in question is divided by the overall N. Thus, the E for the O for Department 1/Safety Glasses is 19×36 divided by $53 = 12.91$. The resulting χ^2 value is 1.19. The df is $(3 - 1)(3 - 1) = 4 \ df$. The assumption by the researcher is that there is no difference between the groups. Since the χ^2 of 1.19 does not exceed the table value of 9.49 for 4 df at the .05 level, the assumption

Table B.5.   Eye Injuries 1995–1996

	Department 1	Department 2	Department 3	Totals
Safety glasses	12	14	10	36
Goggles	4	2	3	9
Face shields	3	2	3	8
	19	18	16	53 = N

O	E	O – E	(O – E)²	$\dfrac{(O-E)^2}{E}$
12	12.91	−.91	.83	.06
4	3.22	.78	.61	.19
3	2.79	.21	.04	.01
14	12.23	1.77	3.13	.26
2	3.10	−1.10	1.21	.39
2	2.71	.29	.08	.03
10	10.86	−.86	.74	.07
3	2.71	.29	.08	.03
3	2.41	.59	.35	.15
53 = N				1.19

$$\chi^2 = \Sigma \frac{(O-E)^2}{E} = 1.19$$

$$df = (k-1)(r-1) = (3-1)(3-1) = 4\ df$$

is correct. The researcher concludes that all three groups have the same safety performance for all three modes of personal protection.

   B.5.3 Corrections. When performing a χ^2 between groups, it is advisable to make a modifying correction if (1) the *df* are very small (such as *df* = 1) and (2) any *E* is less than 10. If both of these circumstances occur, each *O – E* must be decreased by a value of .5, and the χ^2 is then calculated in the usual manner. Note that this correction is not necessary for a 2 × 2 design when using the special formula described in Section B.5.1 because this formula already contains the correction.

Appendix **C**

Categories

SERVICE NO. (NSC)
▶ 1-9 _____
CASE OR FILE NO.
▶ 10-15 _____

SUPPLEMENTARY RECORD OF
OCCUPATIONAL INJURIES AND ILLNESSES

OSHA No. 101   NSC revision
(Meets OSHA requirements
when Instruction 1. has
been followed.)

THIS REPORT IS

▶ 16. 1 ☐ First report   2 ☐ Revised report

EMPLOYER

1. NAME _____

2. MAIL ADDRESS _____

3. LOCATION, if different
from mail address _____

INJURED OR ILL EMPLOYEE

4. NAME _____

SOCIAL SECURITY NO. _____

▶ EMPLOYEE NO.   17-26 _____

5. HOME ADDRESS _____

▶ 6. AGE   27-28 _____

▶ 7. SEX   29. 1 ☐ Male   2 ☐ Female

▶ 8. OCCUPATION (specify) _____

30-31. 01 ☐ Manager, official, proprietor
02 ☐ Professional, technical
03 ☐ Foreman, supervisor
04 ☐ Sales worker
05 ☐ Clerical worker
06 ☐ Craftsman—construction
07 ☐ Craftsman—other
08 ☐ Machinist
09 ☐ Mechanic
10 ☐ Operative (production worker)
11 ☐ Motor vehicle driver
12 ☐ Laborer
13 ☐ Service worker
14 ☐ Agricultural worker
15 ☐ Other
16 ☐ Unknown

9. DEPARTMENT _____
(Enter the name of department or division in which
the injured person is regularly employed.)

CLASSIFICATION OF CASE

A. INJURY OR ILLNESS (see code on Log, OSHA No. 100)

▶ 32. 1 ☐ Injury (10)
2 ☐ Occupational skin disease or disorder (21)
3 ☐ Dust disease of the lungs (pneumoconioses) (22)
4 ☐ Respiratory conditions due to toxic agents (23)
5 ☐ Poisoning (systemic effects of toxic materials) (24)
6 ☐ Disorder due to physical agents
(other than toxic materials) (25)
7 ☐ Disorder due to repeated trauma (26)
8 ☐ All other occupational illnesses (29)

B. EXTENT OF INJURY OR ILLNESS

▶ 33. 1 ☐ Fatality
2 ☐ Lost workday case
3 ☐ Nonfatal case without lost workdays

▶ C. Number of workdays lost   34-36 _____

D. Permanently transferred or terminated

▶ 37. 1 ☐ Yes   2 ☐ No

INSTRUCTIONS

1. Type or print the narrative where requested.
2. Check the one box which most clearly describes
each narrative statement.
3. See also original OSHA No. 101 for more details.
4. Complete form in duplicate. Retain original.
Mail duplicate to: National Safety Council,
425 N. Michigan Ave., Chicago IL 60611.

THE ACCIDENT OR EXPOSURE TO OCCUPATIONAL ILLNESS

10. PLACE OF ACCIDENT
OR EXPOSURE
(mail address) _____

11. WHERE DID ACCIDENT OR EXPOSURE OCCUR?
a. On employer premises

▶ 38. 1 ☐ Yes   2 ☐ No   3 ☐ Unknown

b. Place (specify) _____

▶ 39-40. 01 ☐ Office
02 ☐ Plant, mill
03 ☐ Shipping, receiving, warehouse
04 ☐ Maintenance shop
05 ☐ General or public area of employer premises
(corridor, washroom, lunchroom, parking lot, etc.)
06 ☐ Retail establishment
(store, restaurant, gasoline station, etc.)
07 ☐ Farm
08 ☐ Motor vehicle accident
09 ☐ Other
10 ☐ Unknown

12. WHAT WAS THE EMPLOYEE DOING WHEN INJURED? (Be specific)

_____
_____
_____

a. Task performed at time of accident

▶ 41-42. 01 ☐ Operating machine
02 ☐ Operating hand tool (power or nonpower)
03 ☐ Materials handling
04 ☐ Maintenance & repair—machinery
05 ☐ Maintenance & repair—building & equipment
06 ☐ Motor vehicle driver, operator or passenger
07 ☐ Office and sales tasks, except above
08 ☐ Service tasks, except above
09 ☐ Other
10 ☐ Not performing task
11 ☐ Unknown

b. Activity at time of accident

▶ 43-44. 01 ☐ Climbing
02 ☐ Driving
03 ☐ Jumping
04 ☐ Kneeling
05 ☐ Lying down
06 ☐ Lifting
07 ☐ Reaching, stretching
08 ☐ Riding
09 ☐ Running
10 ☐ Sitting
11 ☐ Standing
12 ☐ Walking
13 ☐ Other
14 ☐ Unknown

Figure C.1 Supplementary Record of Occupational Injuries and Illnesses

13. **HOW DID THE ACCIDENT OCCUR?** (Describe fully the events)

_____
_____
_____
_____

a. AGENCY. (Object or substance involved)

ACCIDENT AGENCY (1st column). The first object or substance involved in accident sequence.

INJURY AGENCY (2nd column). The agency inflicting the injury. See also section 15.

(Example: Worker fell from ladder and struck head on machine. Check "Ladder" under accident and check "Machine" under injury.)

ACCIDENT	INJURY	(Check one box in each column)
45-46. 01 ☐	47-48. 01 ☐	Machine
02 ☐	02 ☐	Conveyor, elevator, hoist
03 ☐	03 ☐	Vehicle
04 ☐	04 ☐	Electrical apparatus
05 ☐	05 ☐	Hand tool
06 ☐	06 ☐	Chemical
07 ☐	07 ☐	Working surface, bench, table, etc.
08 ☐	08 ☐	Floor, walking surface
09 ☐	09 ☐	Bricks, rocks, stones
10 ☐	10 ☐	Box, barrel, container (empty or full)
11 ☐	11 ☐	Door, window, etc.
12 ☐	12 ☐	Ladder
13 ☐	13 ☐	Lumber, woodworking materials
14 ☐	14 ☐	Metal
15 ☐	15 ☐	Stairway, steps
16 ☐	16 ☐	Other
17 ☐	17 ☐	Unknown
18 ☐	18 ☐	None

b. ACCIDENT TYPE. (First event in the accident sequence)

49-50. 01 ☐ Fall from elevation
02 ☐ Fall on same level
03 ☐ Struck against
04 ☐ Struck by
05 ☐ Caught in, under or between
06 ☐ Rubbed or abraded
07 ☐ Bodily reaction
08 ☐ Overexertion
09 ☐ Contact with electrical current
10 ☐ Contact with temperature extremes
11 ☐ Contact with radiations, caustics, toxic and noxious substances
12 ☐ Public transportation accident
13 ☐ Motor vehicle accident
14 ☐ Other
15 ☐ Unknown

This space may be used for additional information.

OCCUPATIONAL INJURY OR ILLNESS

14. **DESCRIBE THE INJURY OR ILLNESS** in detail and indicate the part of the body affected.

_____
_____
_____

a. NATURE OF INJURY OR ILLNESS. (Check most serious one)

51-52. 01 ☐ Amputation
02 ☐ Burn and scald (heat)
03 ☐ Burn (chemical)
04 ☐ Concussion
05 ☐ Crushing injury
06 ☐ Cut, laceration, puncture, abrasion
07 ☐ Fracture
08 ☐ Hernia
09 ☐ Bruise, contusion
10 ☐ Occupational illness
11 ☐ Sprain, strain
12 ☐ Other

b. PART OF BODY. (Check most serious one)

53-54. 01 ☐ Eyes
02 ☐ Head, face, neck
03 ☐ Back
04 ☐ Trunk (except back, internal)
05 ☐ Arm
06 ☐ Hand and wrist
07 ☐ Fingers
08 ☐ Leg
09 ☐ Feet and ankles
10 ☐ Toes
11 ☐ Internal and other

15. **NAME THE OBJECT OR SUBSTANCE WHICH DIRECTLY INJURED THE EMPLOYEE.** Also check one box in injury column under 13a.

_____
_____
_____

16. **DATE OF INJURY OR INITIAL DIAGNOSIS OF OCCUPATIONAL ILLNESS.**

a. MONTH

55-56. 01 ☐ Jan.   07 ☐ July
02 ☐ Feb.   08 ☐ Aug.
03 ☐ March   09 ☐ Sept.
04 ☐ April   10 ☐ Oct.
05 ☐ May   11 ☐ Nov.
06 ☐ June   12 ☐ Dec.

b. DATE OF MONTH   57-58 _____

17. **DID EMPLOYEE DIE?**

59, 1 ☐ Yes   Date of Death _____
2 ☐ No

OTHER

18. **NAME AND ADDRESS OF PHYSICIAN** _____

19. **IF HOSPITALIZED, NAME AND ADDRESS OF HOSPITAL** _____

_____

DATE OF REPORT _____
PREPARED BY _____
OFFICIAL POSITION _____

Figure C.1 Supplementary Record of Occupational Injuries and Illnesses—*Continued*

Tables

Table D.1. Random Numbers

51772	74640	42331	29044	46621	62898	93582	04186	19640	87056
24033	23491	83587	06568	21960	21387	76105	10863	97453	90581
45939	60173	52078	25424	11645	55870	56974	37428	93507	94271
30586	02133	75797	45406	31041	86707	12973	17169	88116	42187
03585	79353	81938	82322	96799	85659	36081	50884	14070	74950
64937	03355	95863	20790	65304	55189	00745	65253	11822	15804
15630	64759	51135	98527	62586	41889	25439	88036	24034	67283
09448	56301	57683	30277	94623	85418	68829	06652	41982	49159
21631	91157	77331	60710	52290	16835	48653	71590	16159	14676
91097	17480	29414	06829	87843	21895	27279	47152	35683	47280
50532	25496	95652	42457	73547	76552	50020	24819	52984	76168
07136	40876	79971	54195	25708	51817	36732	72484	94923	75936
27989	64728	10744	08396	56242	90985	28868	99431	50995	20507
85184	73949	36601	46253	00477	25234	09908	36574	72139	70185
54398	21154	97810	36764	32869	11785	55261	59009	38714	38723
65544	34371	09591	07839	58892	92843	72828	91341	84821	63886
08263	65952	85762	64236	39238	18776	84303	99247	46149	03229
39817	67906	48263	16057	81812	15815	63700	85915	19219	45943
62257	04077	79443	95203	02479	30763	92486	54083	23631	05825
53298	90276	62545	21944	16530	03878	07516	95715	02526	33537

Sampling Procedure:

1. Decide on the sample size.
2. Starting with 0 or 00 or 000 etc. (depending on the size of the group) serially assign a unique identification number to each group member. This may not be necessary if each member already has a unique number such as a Social Security number, payroll number, etc.
3. Starting with the uppermost and leftmost digit in the table, read from left to right across the entire table. After reaching the extreme right hand digit of the row, go to the leftmost digit of the next row below. Readings should be in multiples of 1 or 2 or 3 etc. depending on the size of the group. When a reading coincides with the identifying number of a member, place that member into the sample.
4. To be particularly correct, if a member's number comes up more than once, then the member is placed into the sample more than once. This is referred to as sampling with replacement.

Example:

1. A sample of 10 must be drawn from a corporation of 106 employees.
2. The employees are arbitrarily assigned numbers serially: 000, 001, 002, . . ., up to 106.
3. Starting with the uppermost and leftmost digit in the table and reading across, the sample members would be 042, 058, 056, 063, 104, 064, 033, 045, 034, 027.

Table D.2.   Values of *t* at the 5% and 1% Levels of Significance

Degrees of Freedom (*df*)	5%	1%	Degrees of Freedom (*df*)	5%	1%
1	12.706	63.657	21	2.080	2.831
2	4.303	9.925	22	2.074	2.819
3	3.182	5.841	23	2.069	2.807
4	2.776	4.604	24	2.064	2.797
5	2.571	4.032	25	2.060	2.787
6	2.447	3.707	26	2.056	2.779
7	2.365	3.499	27	2.052	2.771
8	2.306	3.355	28	2.048	2.763
9	2.262	3.250	29	2.045	2.756
10	2.228	3.169	30	2.042	2.750
11	2.201	3.106	35	2.030	2.724
12	2.179	3.055	40	2.021	2.704
13	2.160	3.012	45	2.014	2.690
14	2.145	2.977	50	2.008	2.678
15	2.131	2.947			
16	2.120	2.921	60	2.000	2.660
17	2.110	2.898	70	1.994	2.648
18	2.101	2.878	80	1.990	2.638
19	2.093	2.861	90	1.987	2.632
20	2.086	2.845	100	1.984	2.626
			∞	1.960	2.576

Table D.3. Values of p (Rank-Order Correlation Coefficient) at the 5% and 1% Levels of Significance

N	5%	1%	N	5%	1%
5	1.000	–	16	.506	.665
6	.886	1.000	18	.475	.625
7	.786	.929	20	.450	.591
8	.738	.881	22	.428	.562
9	.683	.833	24	.409	.537
10	.648	.794	26	.392	.515
12	.591	.777	28	.377	.496
14	.544	.714	30	.364	.478

Table D.4. Values of Chi Square (χ^2) at the 5% and 1% Levels of Significance

Degrees of Freedom (df)	5%	1%	Degrees of Freedom (df)	5%	1%
1	3.84	6.64	16	26.30	32.00
2	5.99	9.21	17	27.59	33.41
3	7.82	11.34	18	28.87	34.80
4	9.49	13.28	19	30.14	36.19
5	11.07	15.09	20	31.41	37.57
6	12.59	16.81	21	32.67	38.93
7	14.07	18.48	22	33.92	40.29
8	15.51	20.09	23	35.17	41.64
9	16.92	21.67	24	36.42	42.98
10	18.31	23.21	25	37.65	44.31
11	19.68	24.72	26	38.88	45.64
12	21.03	26.22	27	40.11	46.96
13	22.36	27.69	28	41.34	48.28
14	23.68	29.14	29	42.56	49.59
15	25.00	30.58	30	43.77	50.89

Appendix E

Solutions to Problems

2.1 a. 16 males, 33⅓%
          32 females, 66⅔%
     b. Young (18–34 years)   16, 33⅓%, rounded to 33%
         Middle-aged (35–51)   18, 37½%, rounded to 38%
         Older (52+)   14, 29.2%, rounded to 29%
     c. Range: 20 to 61 years
         Mean: 43.6, rounded to 44 years
         Median: 48 years
     d. Female range: 39 to 61 years
         Female mean: 51.3, rounded to 51 years
         Female median: 51 years
     e. Male range: 20 to 33 years
         Male mean: 28.25, rounded to 28 years
         Male median: 28 years
2.2 May also observe sex, estimated age, department, trade, type of PPE, time of day.
2.3 a. Leg(s): 2
         Face/head: 5
         Fingers: 20
         Eyes: 38
         Other: 42
     b. "Other" category is too large. Try to break it down into smaller categories.
3.1 Chi square (χ^2) = 1.05, not significant

3.2 χ^2 = 10.97, significant
3.3 χ^2 = 21.79, significant
4.1 χ^2 = 122.24, significant
4.2 χ^2 = 4.33, significant
4.3 χ^2 = .125, not significant
4.4 χ^2 Table:

Number of Workers Injured	Number of Workers Not Injured
55	1,024
20	953

χ^2 = 12.59, significant

5.1 χ^2, wing A = .16, not significant
No need to calculate wing B.
Conclusion: Using the lifting devices had no effect on wing A.
5.2 East Coast χ^2 = 6.02, significant
West Coast χ^2 = .09, not significant
Conclusion: The bonus plan had a beneficial effect on the East Coast.
5.3 Line A χ^2 = 24.24, significant
Line B χ^2 = .004, not significant
Conclusion: The protective clothing has increased productivity in line A.
6.1 Correlation of −.78, negative, strong
Significant at .05 and .01 levels
Conclusion: Excessively rapid work leads to more errors.
6.2 Correlation of +.882, positive, strong
Significant at .05 and .01 levels
Conclusion: Departments remain consistently good or consistently bad.
6.3 Correlation of −.21, negative, weak
Not significant, the correlation should be disregarded.
7.1 H = 26.25 corrected to 26.78. At 1 *df*, significant at .05.
7.2 H = 5.55 corrected to 5.56. At 1 *df*, significant at .05.
7.3 H = 19.12 corrected to 19.15. At 1 *df*, significant at .05.
7.4 H = 17.41, no correction needed. At 2 *df*, significant at .05.
8.1 Mean = 202.3
Confidence interval at .05 = 198.0 − 206.6
Confidence interval at .01 = 196.3 − 208.3

8.2 Mean of simple reaction time = 202.3
Confidence interval at .05 = 198.0 − 206.6
Confidence interval at .01 = 196.3 − 208.3
Mean of choice reaction time = 308.75
Confidence interval at .05 = 304.6 − 313.0
Confidence interval at .01 = 302.8 − 314.7
$t = 37.09$, significant at .05 and .01
Conclusion: Choice reaction time is slower than simple reaction time.

8.3 Mean of new section = .47
Confidence interval at .05 = .43 − .50
Mean of old section = .69
Confidence interval at .05 = .66 − .72
$t = 17.48$, significant at .05 and .01
Conclusion: Old section walkways are safer.

9.7 Incidence rate = 13.84
Severity rate = 63.95
Incidence rate = 69.19
Severity rate = 319.74

9.8 Comparison result = 2.77, significant at .05 and .01

9.9 Comparison result = 2.23, significant at .05
$\chi^2 = .178$
The distribution is 50:50.

10.2 $\chi^2 = 1.56$
There is no difference between the groups.

10.3 $\chi^2 = 1.12$
There is no difference between the groups.

10.4 $\chi^2 = .021$
The distribution is 25:75. This will occur in the future.

10.5 $\chi^2 = 7.35$. Monday is no different than other days.

10.6 $\chi^2 = 1.74$. The performance is no different from the others.

11.2 Suspected employees: #230 with χ^2 of .5
                                            #203 with χ^2 of .62
                                            #95 with χ^2 of 1.9
Accident repeaters: None
Employee #230 appears to be a repeater but is not.

12.1 Upper control limit (rounded) = 47
Lower control limit (rounded) = 23

12.2 Upper control limit = 3.3
Lower control limit = 2.2

12.3 Upper control limit (rounded) = 39.7
Lower control limit (rounded) = 34.9

12.4 $151,600 (present worth of savings) > $100,000 (proposed expenditure). This is a worthwhile investment.

12.5 $17,800 (present worth of savings) < $25,000 (proposed expenditure). This is not a worthwhile investment.

Index